From Ancient Sounds to Modern Stages: The Evolution of the Sheng

Mathews

TABLE OF CONTENTS

CHAPTER I. INTRODUCTION...1
    Overview and Background..2
    Genesis of the Instrument...3
    Sheng Performance...6
    Sheng Musicians and Culture...8
    Survey of the Literature..9
    Source Materials..16
    Methodology..17
    Conclusion...18

CHAPTER II. MORPHOLOGY OF THE TRADITIONAL SHENG....................20
    Sheng in Ancient China Before 1949..20
    Characteristics of the Sheng..28
        Wind Chest...28
        Sheng Pipes..30
        Sheng Reeds...32
        Blow Pipe...33
        Tuning System..34
        Materials...36

CHAPTER III. THE EVOLUTION OF THE MODERN SHENG AFTER 1949...............38
    Reformation of National Instruments in People's Republic of China until 1953.........38
        End of the Qing Dynasty, the Last Chinese Empire (1840–1949).....................38
        People's of Republic of China and Influence of the Soviet Union (1949–1953)......43
        Beginnings of Instrument Reform (1954–1966)...46
        The Cultural Revolution (1966–1976)..49
        The Rapid Development of the Sheng after 1976..52
    New Morphologies of the Sheng since 1949..57

CHAPTER IV. GLOBALIZATION AND NEW ROLES FOR TRADITIONAL AND MODERN SHENG PERFORMANCE..79

    Social and Musical Roles of Traditional Sheng Before 1976...79

        Playing Court Music..79

        Playing Sacred Occasions...81

        Playing for Folk Activities..83

        Playing Traditional Opera...83

    From Traditional Ensembles to Conservatory Style..84

    Further Performance Roles for Sheng in the Globalized World.....................................90

CHAPTER V. CONCLUSION...95

APPENDIXES..100

CHAPTER I

INTRODUCTION

The *sheng*, is a Chinese traditional, interruptive free-reed aerophone, consisting of a blow-pipe, bowl, and vertical resonating tubes. The instrument originated in the fourteenth century BCE, and the structure of its design kept changing for some 3200 years. Especially in recent decades, this instrument has been significantly modified. Since the establishment of the People's Republic of China in 1949, the structure of Chinese society has undergone profound shifts. Influenced by political, economic, cultural, and other social factors, the development of Chinese music in general has also shown different orientations with the times.

Other than playing the sheng in traditional ensembles as an accompaniment instrument, Chinese musicians are trying to bring sheng to new stages, such as accompanying in national orchestras, soloing in recitals, and new music ensembles featuring instruments from other countries. In order to adapt to these differing contexts, new demands have been placed on this traditional instrument, such as a lighter weight, more solid construction, broader range, and a new arrangement of pipes for more efficient fingering. Moreover, new materials and processes have been applied to the production of the sheng. The shape of the instrument has also been adapted to meet these demands. As a result, new components and even new kinds of sheng have been developed. This thesis addresses the origin and development of different kinds of sheng in China, and the influence of important social and political movements on Chinese music and musicians. I also explore and analyze the motivations behind the changes of the sheng to demonstrate how Chinese people's views and expectations changed regarding this traditional Chinese musical instrument from 1949 to 2018.

Overview and Background

In this thesis, I discuss the evolution of the sheng in mainland China from 1949 to 2018. Two motivations contributed to my research. For over ten years, I have been a student of sheng performance, studying various types of sheng, including the traditional sheng, Henan square sheng, modern sheng, and bass keyboard sheng. I have taught sheng classes in Nanjing Normal University for two years. With my experience learning and teaching sheng I developed deeper understandings about the instrument's evolution. Through the study of ethnomusicology in the United States I learned necessary knowledge and skills to approach this subject, and, more importantly, a fresh perspective from which to understand music as a product of human behavior. Thus, I will examine the sheng's evolution through the lens of the changes made to the construction of the instrument, the functions fulfilled by such changes, the musicians' intentions in asking for such functions, and the cultural and social factors that created such intentions. As the entire history of the sheng's evolution goes beyond the scope of this thesis, I limit the focus to the years of 1949 to 2018, which corresponds to the establishment of the Chinese communist government. This important political transition deeply influenced subsequent cultural and political ideologies of Chinese society until the current decade.

Genesis of the Instrument

The sheng, also known as the Chinese mouth organ, is a Chinese mouth-blown, free-reed instrument, prevalent among Han people living in north and central-eastern China. A traditional sheng is generally constructed from a bowl-shaped wind chest, with a blow-pipe extending out from one side (see figure 1).

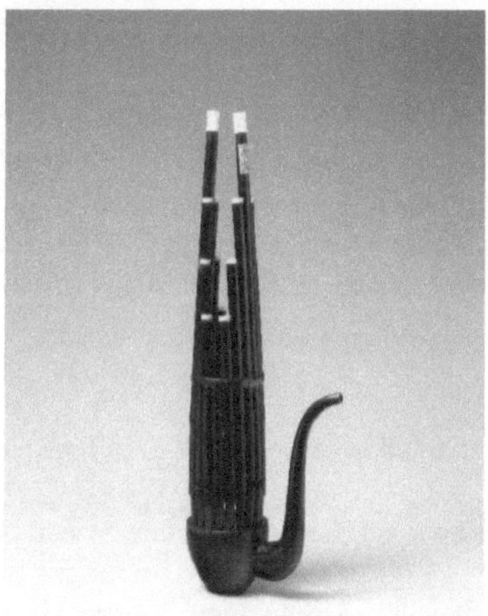

Figure 1. A Traditional Sheng in Qing Dynasty. From the Metropolitan Museum of Art.[1]

Multiple pipes (13, 15, 17, or more) with free reeds at the lower end are vertically inserted through the flat upper surface of the wind chest. The pipes of varying graded-lengths are arranged in two opposite triangular shapes to symbolize the folded wings of a phoenix bird. Most pipes on a traditional sheng sound by covering a small finger hole above the surface of the wind chest. However, some muted pipes without reeds and finger holes remain in the design of the instrument to maintain the instrument's phoenix-like appearance.

Each of the pipes on a sheng plays one pitch in performance, either when inhaling or exhaling. The pitch of a pipe is determined by two dependent factors. The first factor is an air column in the pipe, measured by length and created by an upper aperture cut through the pipe's inner wall. The second factor is the reed, shaped as a rectangular tongue, cut from and seated flush with the frame. The reed is secured with wax at the bottom of the pipe and determines the vibrating pitch frequency. Due to the difficulty of changing the length of the air column or the size or thickness of the reed for tuning, a red wax dot (a mixture of wax, cinnabar, and honey) was invented no later than the Ming Dynasty (1368—1644 CE). Sheng players applied the wax near the sheng's vibrating end onto the reed tongue for finer adjustments of pitch, accomplished by changing the size of the red wax dot. When open, the finger holes on pipes above the wind chest's flat surface function to break the air columns that reinforce the reed vibration, hence, little sound is made. In contrast, a closed finger hole seals the air column so the reed vibration is reinforced, and a clear sound can be made.

The materials used in sheng construction have evolved over time. The earliest sheng, dating from around the sixth century BCE, have gourd or wood wind chests and bamboo pipes. Their reeds are made of copper, bamboo, or similar plant material. The tomb of Marquis Yi of Zeng (曾侯乙墓，zēng hóu yǐ mù), in Hebei province, dates to 433 BCE. It contains five small mouth organs with gourd wind chests, bamboo pipes, and bamboo reeds.[2] The Han Dynasty (206 BCE—220 CE) tombs of the Mawangdui archeological site in Hunan

---

[2] Wujian Jiang, "*Zenghou Yi Mu Sheng de Yinluxue Yanjiu* [Restoration of the Sheng Discovered in the Tomb of Marquis Yi]," *Huangzhong* 3 (1998): 36.

province (马王堆汉墓，mǎ wáng duī hàn mù), dating to the second century BCE, contain two large *yu* (Chinese wind instruments similar to the sheng) with wooden wind chests and bamboo pipes, one of which uses metal reeds. Some silver metal beads were also found attached to the reeds and are considered evidence of the origins of the red wax tuning idea. Nowadays, sheng craftsmen use not only traditional materials, like bamboo and wood, but also materials such as alloys and plastics. Metal wind chests, metal levers, plastic buttons, stainless-steel amplifying tubes, electroplated coatings, and many other modern components are often seen on the sheng.

The tuning system of a traditional sheng is essentially diatonic, while sometimes a seventh and fourth interval may be added. For example, if a sheng is tuned in D major, other than essential pipes playing D, E, F-sharp, A, and B, an extra pipe of C-sharp maybe added. The fingering is arranged for the pentatonic scale regardless of the instrument's specific tonality (i.e., D major, E-flat minor, C major). Nowadays, in order to produce louder sounds and more pitches, a traditional sheng may also have amplifying metal tubes and levered keys. The Henan traditional sheng, or *fangsheng*, prevalent in the Henan province and the southwestern part of the Shandong province, is an exception to the round wind chest definition, with rectangular wind chest and pipes arranged in three parallel rows.

As the modernization movement in Chinese traditional music and musical instruments unfolds in the beginning of the twentieth century, new kinds of sheng emerged in China in the 1950s. Two essential differences distinguish these new sheng from the traditional sheng: 1) all the new sheng are tuned chromatically, not diatonically, and 2) levered keys that close the pipe ends are introduced, and the reed vibration is no longer controlled by finger holes alone. Some kinds of new sheng have applied the rectangular wind chest and arranged pipes in three parallel rows. However, in order to produce lower pitches and more stable sounds with an alto, tenor, or bass sheng, artisans and musicians have increased the volume of its resonating tubes, making it a large metal-made, standing organ-like instrument, used with or without pedals (see figure 2).

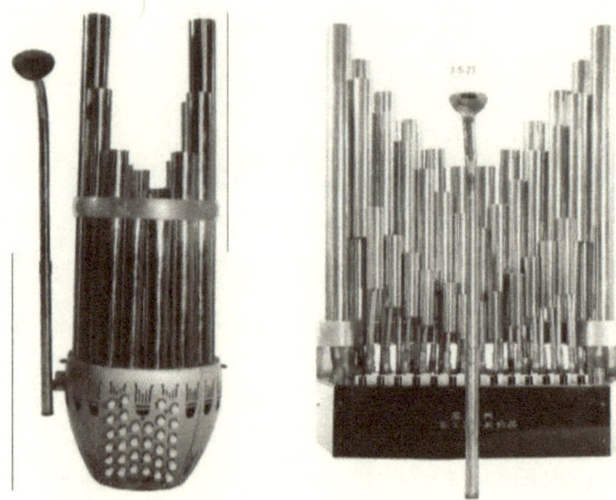

Figure 2. Bass Holding Sheng (left) and Alto Pai Sheng (right). From *Zhongguo Yueqi Tujian* [Chinese Musical Instruments Illustration].[3]

Sheng Performance

The social and cultural context, including prevailing or controversial issues of the day, shape the mainstream aesthetic and ways of expression for sheng performance, as well as popular sheng repertoire. Audience and popular perception of instruments determines the sheng's role in different ensembles. For example, a traditional sheng is often seen for

---

[3] Dongsheng Liu. *Zhongguo Yueqi Tujian* [Chinese Musical Instruments Illustration] (Qingdao: Shandong Education Press, 1992), 160-161.

entertainment occasions and folk and ritual ceremonies. In *kunqu* (a traditional genre of

Chinese opera) and *sizhu* ensembles (a traditional genre of Chinese ensemble), the sheng's

role is confined to an important accompaniment, following the *dizi*'s (a traditional Chinese

flute) melody. However, in ritual occasions, such as Buddhist and Taoist ceremonies, the

sheng is a leading instrument, providing the structure of the music and controlling the pace of

the ritual.

The traditional use of sheng changed as the Chinese Communist Party's (CCP)

nationwide, anti-feudal, and "anti-superstition" social movements unfolded in the 1950s.

Instead of entertainment music or religious ritual performance, traditional musicians were

encouraged to compose and perform new music for labor and peasants, thus complying with

the government's political agenda. By contrast, the collectivism and command economy in

China limited the business market for traditional ensembles.[4] As a result, some sheng players

started to play as soloists, thus, although used as an accompaniment instrument for hundreds

of years, the traditional sheng became known as a solo instrument with unique polyphonic

features.

Though it can be used to play solo pieces, the newly designed modern sheng are

more often heard in an ensemble, e.g., as part of the Chinese national orchestras in concert

---

[4] Collectivism is any of several types of social organization in which the individual is seen as being subordinate to a social collectivity, such as a state, a nation, a race, or a social class. Collectivism may be contrasted with individualism, in which the rights and interests of the individual are emphasized; *Encyclopedia Britannica Online*, s.v. "Collectivism," accessed April 22, 2020,; In China, command economy is an economic system in which the means of production are publicly owned and economic activity is controlled by a central authority, which assigns quantitative production goals and allots raw materials to productive enterprises; *Encyclopedia Britannica Online*, s.v. "Command Economy," accessed April 22, 2020, https://www.britannica.com/topic/command-economy.

halls. Chromatic, twenty-four reed and twenty-six reed new sheng were common during the 1950s, but today's models usually have thirty-two to thirty-eight pipes. The modern sheng is essentially a result of the westernization of Chinese traditional music and musical instruments beginning in the early twentieth century.

Deeply influenced by foreign cultures, especially by ideologies from western Europe and the Soviet Union, Chinese intellectuals in the late 19th and early 20th centuries held strong beliefs about industrialization and modernization. Therefore, the equal-tempered tuning system was introduced into China as part of "advanced" western culture and promoted throughout the nation. As a result, the traditional instruments that used the Pythagorean tuning system, sheng included, were modified to fit an equal-tempered tuning system. In national orchestras, as the only polyphonic traditional Chinese instrument, the sheng still plays an important role as both solo and accompaniment instrument.

Sheng Musicians and Culture

Many devoted sheng players, composers, and artisans contributed to the evolution of the sheng, conceiving and achieving improvements on the instrument and its music. Tianquan Hu (b.1934—), a Shandong sheng player, won two gold medals at the sixth "World Festival of Youth and Students," in 1957, in Moscow for his sheng solo performance of *Fenghuangzhanchi* [The Phoenix Unfolds Its Wings]. Hu added a *bawu* pipe (an individual free-reed aerophone, originating in Yunnan Province of southwest China) to his sheng, thus he could play both instruments in a single piece without switching. In the 1980s, Zhenfa

Weng (n.d.), together with Shanping Mu (b.1942—), both from Shandong Conservatory,

invented the 37-reed modern sheng. Since the 1990s, as a technical consultant, Jiangong Lei

(b.1946—), from Nanjing Conservatory, has been working on sheng design with artisans

from the Zhaojiasheng Company in Hebei province. Together they have implemented design

changes to overcome many technical difficulties in modern sheng building and formulated

standards for sheng components, such as reeds, bamboo pipes, and the structure of the bass

keyboard sheng.

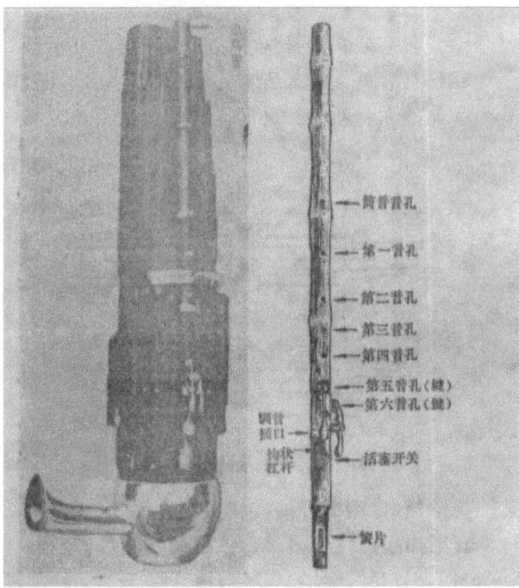

Figure 3. Bawu Sheng (left) and its bawu pipe (right).[5]

Survey of the Literature

Academic study of the sheng began in the 1930s. Among other studies, Yang Yinliu

---

[5] Tianquan Hu, "Bawu Sheng Gaige Zhong De Jige Wenti," [Some Problems I Encountered When Inventing the Bawu Sheng] *Yueqi* 6 (1982): 11-13.

(杨荫浏) (1899—1984) conducted comparative research between two similar instruments—the sheng and the *yu* (an obsolete traditional Chinese interruptive free-reed aerophone, similar to, though larger in size than the sheng)—completing his 1944 article, "Comparative Research of Sheng and Yu," eventually published in *Draft History of Chinese Music* in 1966.[6] In the 1950s, musician Pan Huaisu (潘怀素) (1894—1978); guqin master, Zha Fuxi (查阜西) (1895—1976); pipa (a traditional Chinese lute) master, Yang Dajun (杨大军) (1913—1987); and Yang Yinliu discovered the fifteenth century Ming Dynasty *sheng-guan* repertoires preserved in the Zhihua Temple in Beijing. They launched a restoration project that later reproduced the royal Buddhist ritual music of the Ming Dynasty. In addition, Yang Yinliu spent years in Hebei villages doing fieldwork on the Hebei sheng, making a considerable contribution to the study of sheng. Research results published in the 1950s and related to sheng study include *Zhihuasi Beijing Music,* by Yang Yinliu, and *Suitang Yanyue de Chengli, Dibian yu Liuchuan* [The Formation, Transmutation and Spread of Yan music in Sui and Tang Dynasty], by Pan Huaisu.[7]

In the 1980s, more materials related to the sheng were published. The *Dictionary of Chinese Music* included information about how the sheng is played in bands of traditional Chinese genres, such as *kunqu* opera, *sizhu, liu* (a traditional Chinese opera originating in Shandong region), and *bangzi* (a traditional Chinese opera originating in Hebei region).[8]

---

[6] Yinliu Yang, *Zhongguo Yueqi Shi Gao* [Draft History of Chinese Music] (Beijing: People's Music Publishing House, 1966), 88.

[7] Yinliu Yang, "Interview in Zhihua Temple, Beijing," *Collected Works of Yang Yinliu,* vol. 6 (Nanjing: Jiangsu Literature and Art Publishing House, 2009), 144-267; Huaisu Pan, "The Formation, Transmutation and Spread of Yan music in Sui and Tang Dynasty," *People's Music*, no. 1 (1954), 29-32.

[8] Chinese National Academy of Arts, *Dictionary of Chinese Music* (Beijing People's Music Publishing House, 1984), 331-333.

Fredrick Lau explained the morphology of the instrument, its diffusion and performance in his article "Instrument: Sheng."[9] Jing Weigang (景蔚岗) published the *Notation System for Sheng in Northern Shaanxi*, in1988.[10] Beside these articles, pictures of a sheng from the nineteenth century and pictures of components in detail are displayed on the website of the National Music Museum.[11] Although these articles all describe the sheng as a mouth-blown, free-reed, vertically played aerophone with bamboo pipes and a wind chest made of gourd, wood, or metal, none of them give any information about the impact of historical, cultural, or social functions of this instrument.

Since the 1990s, more research has emerged concerning the music and performance of the sheng. Several studies focus on sheng music and performance within specific traditions of China, including Gao Pei (高沛)'s "The Henan Sheng and its Performing Characteristics," Jing Weigang's "The Clarification of Name and Content of Shanxi's Eight Grand Suites," and Ou Jie (欧杰)'s "Sheng's Performing Skills in Xi'an Chuida Ensembles."[12] These are important for the study of sheng music and performance in each tradition and are also reliable resources for the comparative study of sheng music and performance in different geographic areas and during different eras, thus forming a more dynamic understanding of the evolution of the sheng in China. Nevertheless, these studies offer few, if any, insights on the connection

---

[9] Fredrick Lau, "Instrument: Sheng," *Garland Encyclopedia of World Music: East Asia: China, Japan and Korea*, edited by Rober Provine, vol. 7 (New York: Routledge, 2001), 186-189.

[10] Weigang Jing, "Notation System for Sheng in Northern Shaanxi" (M.A Thesis, Chinese National Academy of Arts, 1988).

[11] "National Music Museum," accessed January 30, 2019, http://collections.nmmusd.org/EasternAsia/2566ChineseSheng/Sheng2566.html.

[12] Pei Gao, "The Henan Sheng and its Performing Characteristics," *Yueqi* 5 (1994): 35-39; Weigang Jing, "The Clarification of Name and Content of Shanxi's Eight Grand Suites," *Music Research* 1 (March 2004): 42-51; Jie Ou, "Sheng's Performing Skills in Xi'an Chuida Ensembles," (M.A. Thesis, Nanjing Conservatory), 2013.

between the traditions of sheng music and local culture.

Thus, most academic research materials related to sheng examine this instrument and its music in a historical or musicological scope. For example, Gao Pei's "The Evolution and Development of the Sheng" addresses the sheng's development from its genesis to the current instrument. The author arranges every detail into a complete story but is not able to convincingly argue all his points with detailed evidence within the space of a short journal article. Gao mentions that the largest sheng from the Han Dynasty has as many as thirty-six pipes, while the smallest sheng from the Qing Dynasty (1644 —1912) has only seven. Gao argues this disparity is the result of the Han people adopting instruments and music genres from other cultures, which has weakened the status and function of the sheng in traditional ensembles. Though this suggestion could be correct, Gao offers no solid evidence, such as with the smallest sheng from the Han Dynasty, the largest sheng from the Qing Dynasty, and other archeological discoveries proving the existence of such a change during the hundreds of years between these two dynasties. Fortunately, Gao clearly describes the storyline of the sheng's evolution and the spread of sheng music and performance, thus helping to focus my research to target how the sheng is commonly used and recognized in Chinese society.

Materials studying the connection of music and Chinese society were also important in this study. Ju Qihong (居其宏), in his "History of Music in New China (1949–2000)," published in 2002, gives a comprehensive description of the evolution of music traditions in China after 1949.[13] The author focuses on how these traditions changed in the

---

[13] Qihong Ju, *History of Music in New China (1949-2000)* (Wuhan: Hunan Art Publish, 2002).

social, political, and cultural upheaval of 1949 to the year 2000. He addresses social

movements and events as triggers of musical changes, listing significant national political

events, such as the Great Leap Forward (an economic and political campaign by the

Communist Party of China from 1958 to 1962), the Great Chinese Famine (a national-wide

famine from 1959 to 1961), the Great Proletarian Cultural Revolution (a sociopolitical

movement in China from 1966 to 1976), and the Chinese Economic Reform (also known as

the Opening of China, starting in 1978). He explains the connection between the musical

works and the social and political motivations behind their creation. While Ju Qihong

presents Chinese music history in the overview, the sheng as an instrument with particular

cultural and historical meanings is beyond the scope of his work. Although he mentions

genres that include the sheng, e.g., performance of ensembles in village ceremonies, the book

does not serve well as a single study of the sheng and sheng music as it emphasizes cultural

and aesthetic issues. It does, however, work well as a reference for discovering possibly

suppressed or encouraged uses of sheng music.

Some studies discuss the motivations of the development of Chinese music from a

cultural and aesthetic perspective. In *The Cultural Perspective of Chinese Music Aesthetic,*

Jianhua Guan attempts to explain traditional Chinese music as a product of the Chinese

language while asserting that modern Chinese music is "simply an imported western

product."[14] Because of this difference, he claims he always feels a profound inconsistency in

"modernizing" Chinese music. Guan's conclusion supports the pipa scholar Man Wu's

---

[14] Jianhua Guan, *The Cultural Perspective of Chinese Music Aesthetic* (Nanjing: Nanjing Normal University Press, 2013), 144.

argument that "the modernization of Chinese traditional music is a total failure" (中国民乐西洋化，是败笔).[15] However, questions arise with such conclusions; for example, what do these scholars mean by the "profound inconsistency" brought by the modernization of Chinese music? Did this profound inconsistency happen to the sheng as well? In this thesis, I will list and analyze the motives, processes and final results of movements carried out under the guidance of the westernization and modernization of Chinese traditional music starting from the beginning of the twentieth century. Comparing the musicians' original intention and the final results, combined with the analysis of such objections, I discuss my own conclusion on whether the westernization and modernization of the sheng failed.

In addition to published literature, I use iconographic and recorded materials to construct a historical, cultural, and political context that better and more accurately describe an understanding of how music was produced after 1949. For example, the sheng players Wu Wei (吴巍) and Wu Tong (吴彤) represent sheng musicians bringing the sheng to the world's stage. The analysis of their collections reveals both sheng musicians' aesthetic preference in selecting, arranging, and performing pieces. Additionally, these sources demonstrate the influence of culture and society, e.g., the reform and opening policy from 1978 and the following globalization on musicians' ideas about music creation.[16]

---

[15] "Interview: Man Wu," accessed January 31, 2020, https://www.thepaper.cn/news Detail_forward_1657665.

[16] After Mao Zedong's death in October 1976, Hua Guofeng, the paramount leader at that time, together with Ye Jianying and Wang Dongxing arrested the Gang of Four, putting an end to the Cultural Revolution. Economic reforms began during the following "Boluan Fanzheng" [Eliminating chaos and returning to normal] period. In The 3rd Plenary Session of the 11th Central Committee of the Chinese Communist Party in December 1978, Deng Xiaoping replaced Hua Guofeng as the paramount leader of China. Deng implemented a serious of political and economic reforms which are summarized as "reform the domestic system and open-up to the world," or "reform and open policy." Modern China's participation in globalization starts with these reforms.

By reviewing the texts and scores listed above, a map emerges of how sheng research has developed since the 1940s. With the mindset that western civilization is advanced, Chinese intellectuals in the early twentieth century introduced a large number of western civilization materials into China, including western music theory and instruments. As an instrument widely used in traditional ensembles in villages and cities, the sheng was not treated with respect until the 1950s. Only a few musicians kept studying and developing further understanding of Chinese traditional music. Due to a lack of financial support, these musicians primarily worked alone in observing, recording, interpreting, learning, and restoring traditional sheng instruments and music.

The 1960s through the 1970s were significant years in terms of a series of radical political movements launched by the CCP, and subsequent violence, especially the Great Proletarian Cultural Revolution from 1966 to 1976, when musicians could hardly continue their research on the sheng as a traditional instrument. The situation became even more dangerous for music traditions of villages, or of ritual ceremonies, because such traditions were considered outdated, superstitious, or hostile to the new government. Thus, due to Tianquan Hu's success (as noted above) in Moscow, musicians during these decades showed more preference for the sheng's capability and potential as a modern solo instrument than for its traditional use.

But as the Chinese Economic Reform unfolded in China in the 1980s, the Communist Party authorities began to introduce capitalistic market principles into the domestic economic system. As a result, pressures subsided that had prevented scholars from researching sheng music and performance in rural areas or for religious occasions. The

government also began to realize the value of traditional music in preserving traditional

Chinese culture. Thus, in 1992, a project to write an encyclopedia of folk instrument

repertoires from all ethnicities in China, called *The Collection of Chinese Folk Instrumental

Music*, was launched by the Minister of Culture, State Ethnic Affairs Commission and

Chinese Musicians' Association.[17] Sheng repertoire from different genres and regions was

included in this collection. In the 1990s, with increased communication between musicians

from China and countries abroad, Chinese musicians began learning and applying theories

and methods from ethnomusicology, sociology, and other related disciplines into their

research on the sheng.

Source Materials

Data acquired for this thesis comes from both primary and secondary resources,

including photographs, audio and video recordings, newspaper reports, interviews, music

scores, scholarly books, journal articles, theses, dissertations, encyclopedias, fieldwork

reports, lectures, and speeches. They also derive from my personal experiences, observations,

and reflections while learning, teaching and performing on sheng in China and the United

States.

Texts include reports, scores, scholarly books, journal articles, theses,

dissertations, all available from schools, orchestras, public, and school libraries, websites,

---

[17] Lin Li, ed., *The Collection of Chinese Folk Instrumental Music* (Beijing: People's Music
Publishing House, 2005).

book retailers, and online encyclopedias. I accessed many materials, including books, journals, encyclopedias, and scores, from local library systems, such as the Performing Arts Library at Kent State University and the Ohio State Library. I also researched sources from schools in China, for example, the School of Music in Nanjing Normal University, Nanjing Conservatory, National Orchestra of Nanjing City, Shandong Conservatory, Shanghai Conservatory, China Conservatory and Central Conservatory of Music. Staff at these institutions were willing to help me borrow or copy documents and scores. Some books and repertoires were available for purchase on Amazon, Taobao, JD, and other online booksellers. Some of these vendors sell e-books, which are easy to read on mobile devices.

Free access to databases, e.g., the *Garland Encyclopedia of Music* and JSTOR, were provided by Kent State University. Other resources, like the *Encyclopedia Britannica,* were free to access. Varying websites were utilized for other sources such as interviews, newspaper and journal articles, reviews, and criticism. Audio and video resources included portraits of significant musicians, audio and video recordings of recitals and concerts, published CDs and DVDs, as well as pictures and videos taken during fieldwork.

Methodology

The primary methodology for this study was through participant-observation. In my fieldwork, I interviewed many sheng players, composers, and instrument-makers and manufacturers, asking two main research questions: 1) Why did practitioners modify their sheng, and 2) what were the outcomes of these modifications? The complexities of the sheng include its size, weight, appearance on stage, fingerings, sound effects, and difficulty in maintenance. All are important factors to consider when changing its design. When I play Shandong sheng, I find it almost impossible to play the Shanxi repertoire, due to the fingering. For example, the major-second chord common in Shandong repertoire is virtually impossible to finger on a Shanxi-style sheng. Thus, in this thesis, I address the issue of how musicians try to solve such problems. Some musicians add more keys to their instrument, hoping to build a "universal" sheng that fits every repertoire. Others claim that we should learn from Japanese musicians who retained the *shō*, a Japanese mouth organ, in its original form for centuries. Ideally, these musicians believe that a sheng should also be as simple and original as possible.

Such ideas for innovation have resulted from sheng musicians' personal understanding of the music, the instrument, and Chinese culture. I examine their testimonies and demonstrate the connections between the instrumental music of the sheng and society and culture. I turn to my own experience playing and teaching the sheng to understand the intention and function of each development of the instrument and its music.

Conclusion

In this thesis, the core purpose of this research is to examine the functions fulfilled by the sheng in its modification in mainland China since 1949, corresponding to the changes brought by a series of social, cultural, and political movements in Chinese society. Through my examination of the sheng in historical and modern contexts, I illustrate how this instrument, its performance, and repertoire, adapted to new cultural developments. By

analyzing these elements, I provide a more comprehensive narrative of the development of

the sheng and sheng music and how these resulted from changes in Chinese society during

this period from 1949 to today.

CHAPTER II

MORPHOLOGY OF THE TRADITIONAL SHENG

In Chinese culture, as the lead instrument for imperial court ensembles for several hundred years, the sheng is one of the oldest and most highly respected musical instruments. It has been prevalent among Han people living in north and central-eastern China for thousands of years. Certainly, throughout this long history, the sheng has experienced significant developments. Although the focus of this thesis is the morphological development of the sheng from 1949 to 2018, it is necessary to present an overview of the development of the sheng before 1949. I have two reasons to do this: first, as a traditional Chinese musical instrument with a long and continuous history, the cultural and social content and functions in performances of the sheng before 1949 profoundly influenced the development orientation of the sheng after 1949; secondly, although the concept of westernization played a very important role in the morphological development of the sheng from 1949 to 2018, the musicians and the artisans did not ignore the value of eralier traditions. On the contrary, they learned a lot from the structure, materials, techniques and so on from the ancient sheng and applied this knowledge to create new types.

Sheng in Ancient China Before 1949

As one of the oldest Chinese-originated musical instruments, the sheng was recorded in the earliest known form of Chinese writing—the Oracle Bone Script--thus engraved on the oracle bones[18] as "龢" (Figure 4), pronounced "hé," the script indicating

plural bamboo pipes with mouthpieces tied together and arranged according to their pitches.

According to archeological discoveries, the majority of Oracle Bone Script were used for

pyromantic divination during the period of the last nine kings of the Shang Dynasty, dating

from the fourteenth to twelfth centuries BCE. Diviners would submit questions to deities

regarding future weather, crop planting, the fortunes of members of the royal family, military

endeavors, and other similar topics by carving these questions onto a bone or turtle shell in

oracle bone script with a sharp tool, and heat it until the bone or shell cracked. The diviner

would then interpret the pattern of cracks and write the prognostication upon the piece as

well.[19] The discovery of the oracle scripts that represent the sheng indicated that the

prototype of the sheng had already existed in the Shang Dynasty. Although Oracle Bone

Script of "龢" was identified as "types" of sheng in later texts, the first appearance of the

character "笙" for sheng is seen in *Shijing* 《诗经》 [Classics of Poetry] in the Zhou

Dynasty (1100 BCE–256 BCE). [20]

Figure 4. Chinese Oracle Bone script for "sheng" from *Shuowen Jiezi* by Xu Shen （许慎） in Han Dynasty. The left side indicates plural bamboo pipes with mouthpieces tied together and arranged in order based on pitches. The right side indicates a human player.[21]

During the Zhou Dynasty, the sheng was recorded in texts and was the largest played musical instrument at the time. According to the oral tradition of the western Zhou Dynasty (1100 BCE–771 BCE), King Mu of Zhou appreciated music, especially sheng performance.[22] Among the instrumental music programs with sheng works performed often for him were "You Geng" 《由庚,》 "Chong Shan"《崇山,》 "You Yi"《由仪,》 "Nan Gai" 《南陔,》 "Bai Hua" 《白华,》 and "Hua Shu"《华黍.》[23] These sheng pieces were the most popular and are considered the earliest notated sheng works of ancient China.

In addition, starting with the Zhou Dynasty, the classifications of size, shape, number of pipes, and materials of the sheng appear in the historical literature. The *Erya·Shiyue* 《尔雅·释乐》 [Erya·Explain Music] concluded that "The larger size sheng is named *Chao* '巢,' the smaller size sheng is called *He* '和.'"[24] Zheng Xuan, in *Zhouli·Chunguan·Shengshi* 《周礼·春官·笙师》 [Rites of Zhou·Offices of

---

[21] Shen Xu, *Shuowen Jiezi* [Chinese Charactors]. (Beijing: Jiuzhou Press, 2001), 121.

[22] King Mu of Zhou, personal name Ji Man, was the fifth king of the western Zhou Dynasty of ancient China. The dates of his reign are approximately 1026 BCE–922BCE.

[23] Shan Wang, "Sheng De Lishi Yanbian Yu Gaige Tanjiu" [Evolution and Development of Sheng], *National Music*, (March 2017): 7.

[24] *Erya* 《尔雅》 [Erya] is the first surviving Chinese dictionary, dating from the third century BCE.

Spring·Instructor of Sheng], noted that the "Sheng has different numbers of pipes, some of them have thirty-six reeds, some of them have sixteen reeds."[25] Gao Xiu, in *Lüshi Chunqiu* 《吕氏春秋》 [Lüshi Chunqiu], noted that the "*Yu* '竽' is the large-sized sheng and made of gourd. The Yu has thirty-six reeds and the sheng has seventeen reeds."[26] These records in the historical literature show that, since the Zhou Dynasty, musical instruments of the sheng family have been distinguished by their sizes and, in order to classify them - the terms *Chao* "巢, " *Yu* "竽, " *Sheng* "笙, " and *He* "和, " were created. Regrettably, evidence of all the above-mentioned historical sheng materials is found only in texts; no physical sheng certification has been found except for the Han Dynasty sheng.

Among archeological find at the *Zēng Hóu Yǐ Mù* 曾侯乙墓 [Tomb of Marquis Yi of Zeng][27] in Hebei province, the earliest physical evidence of sheng instruments dating to 433 BCE was discovered in 1978. It includes five small mouth organs (see Figure 5).[28] These instruments illustrated the characteristic of the bowl of the sheng as an ellipse based on the natural form of the gourd, with fourteen holes corresponding to fourteen sheng pipes (see Figure 6). At another archaeological site located in Changsha—*Mǎwángduī* 马王堆汉墓 [Mawangdui],[29] two large Yu (竽) were discovered during the excavation of the tombs of

---

[25] *Zhouli* 《周礼》 [Rites of Zhou] is a work on bureaucracy or organizational theory, which is divided into six chapters: 1. Offices of the Heaven, 2. Offices of Earth, 3. Offices of Spring, 4. Offices of Summer, 5. Offices of Autumn, and 6. Offices of Winter; Pei Gao, "Sheng De Yange He Fazhan" [The Development and Standardization of Sheng], *Yueqi*, no.1, (January 1986): 3.

[26] *Lüshi Chunqiu* 《吕氏春秋》 [*Lüshi Chunqiu*] is an encyclopedic Chinese classic text, compiled around 239 BCE.

[27] *Zēng Hóu Yǐ Mù* 曾侯乙墓 [Tomb of Marquis Yi of Zeng] is an archaeological site in leigutun, Zengdu District, Suizhou, Hubei Province, China, dated around 433 BCE.

[28] Wujian Jiang, "Restoration of the Sheng Discovered in the Tomb of Marquis Yi," *Huangzhong* 3 (1998): 36.

[29] *Mǎwángduī* 马王堆汉墓 [Mawangdui] is an archaeological site located in Changsha, Hunan Province, China. The site contains the tombs of three people from the Changsha Kingdom, during the western

western Han Dynasty (206 BCE–9 CE), between 1972 and 1974 (Figure 7). The excavated

Mawangdui sheng not only showed the integrated morphology of the large sheng of the Han

Dynasty, but also the wisdom of its unique design, with the folded pipes in the sheng, which

also provided an idea for the innovation of the future sheng (see Figure 8).

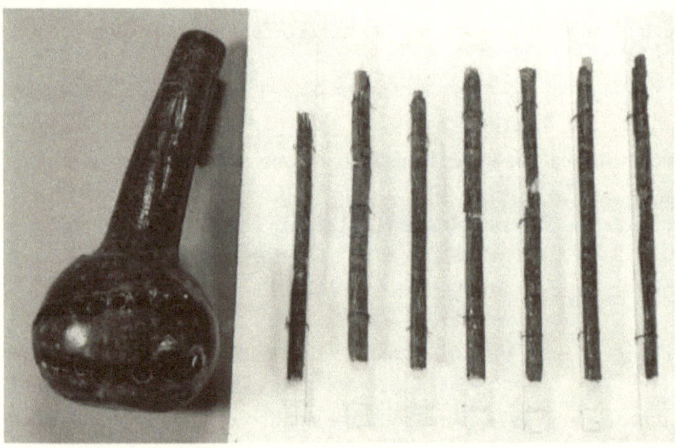

Figure 5. Sheng discovered in the Tomb of Marquis Yi of Zeng, gourd made bowl on the left and bamboo pipes on the right.[30]

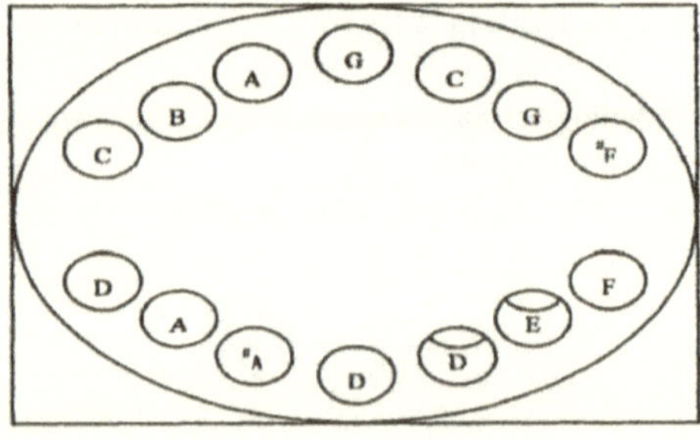

Figure 6. Ellipse-shaped gourd bowl and positions of pitches of the sheng in the Tomb of Marquis Yi of Zeng.[31]

Han Dynasty (206 BCE–9CE).

    [30] Chinese National Academy of Art, *Illustrated Handbook of Chinese Musical Instruments*, (Jinan: Shandong Education Press, 1992), 164.

    [31] Ibid. 164.

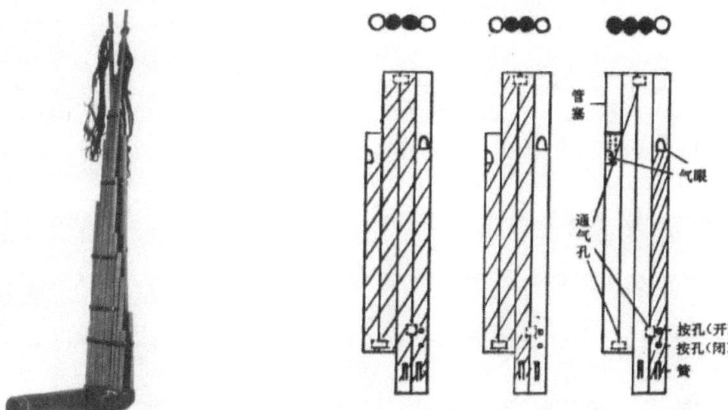

Figure 7. Yu discovered in Mawangdui.[32]   Figure 8. Structure of the folded pipe in Mawangdui yu.[33]

According to the archeological excavated findings, the sheng was not only popular

in imperial court music, but also became prevalent among folk people (approximately 770

BCE–221BCE) during the Spring and Autumn periods. At that time, the Emperors of Zhou

and the central court gradually lost power to local state leaders. Sheng players from royal

court ensembles were employed by both local aristocrats and ordinary people, bringing the

sheng and performance skills to a wider population. From the Warring States Period to the

Han Dynasty (475BCE–220CE), the sheng always played the role of lead instrument in

musical ensembles.

During the Sui and Tang Dynasties (581CE–907CE), the thirteen-, seventeen-, and

nineteen-reed sheng were very popular in both court music and folk music ensembles. In

---

[32] Chinese National Academy of Art, *Illustrated Handbook of Chinese Musical Instruments*, (Jinan: Shandong Education Press, 1992), 165.

[33] Ibid. 165

particular, the concept of "Yiguan sheng" 义管笙 [sheng with support pipes] became fashionable in the sheng community, which marked the outset of sheng players paying attention to the instrumental function of modulation, and also the maturity of the sheng morphology. During the Tang Dynasty (618 CE–907CE), at the peak of sheng popularity, the thirty-six-reed sheng appeared. With the sheng's maturity of morphology it became a representative Chinese instrument, and was introduced to Japan during the Tang Dynasty, with three types of yu and three of sheng, which remain preserved in the Shosoin imperial repository in Nara, Japan today (Figure 9). Moreover, the performance skills of the sheng in the Tang Dynasty became also highly developed, along with the upgrading of its instrumental function. Many sheng masters emerged at this time, such as Yuchi Zhang 尉迟章, Fan Hangong 范汉恭, Fan Baoshuai 范宝帅, and Meng Cairen 孟才人.

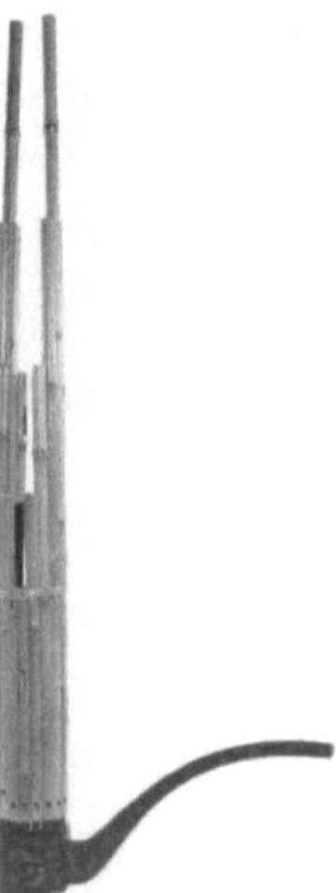

Figure 9. Yu from Tang Dynasty, preserved in Shosoyin, Nara, Japan.[34]

From the Tang Dynasty onwards, musical instruments such as the pipa began to gradually gain favor; therefore, although the sheng still played an important role, it passed its peak period and began to gradually decline in popularity.

In the Song Dynasty (960 CE–1279CE), court musicians rigidly standardized multiple types of sheng developed by their predecessors into a unified sheng of nineteen pipes and nineteen reeds (or a seventeen-pipe sheng with two false pipes), thus classified these types of sheng into three categories: *Chao Sheng* 巢笙, *He Sheng* 和笙, and *Yu Sheng* 竽笙. The Chao sheng is ideally tuned in C major, the He sheng is tuned in F major, and the Yu sheng is tuned in E♭ major respectively.[35] The practice of tunings, which further standardized the sheng, but also imposed more restrictions on its declining already development, thus inhibiting more developments of the instrument.

In the Yuan, Ming, and Qing Dynasties (1271 CE–1912 CE), the lead and solo instrument status of the sheng fell away, and it became a secondary, accompaniment instrument. The sheng of thirty-six pipes, reflecting the best technical developments thus far, appeared in the Tang Dynasty, existing briefly before being reduced to seventeen or fourteen pipes, and later further reduced to as few as nine pipes. Since the Tang Dynasty, the number of sheng pipes continued to decrease as the instrument itself became less popular; however, in late Qing Dynasty, folk musicians from the Henan area developed a sheng with a square-shaped wind chest, into which fourteen pipes were arranged in three parallel rows. This style

---

[35] Yang Yinliu's calculation results. Yang Yinliu 杨荫浏 (1899-1984) was a musicologist from the People's Republic of China.

of sheng offered musicians ideas for forming the modern sheng decades later.

Characteristics of the Sheng

On examination of the different types of sheng in ancient China shows that the shape of the instrument has constantly changed throughout history since its invention. Therefore, rather than viewing the sheng as an instrument with a certain, unchanging form, it is better viewed as a family of instruments. Therefore to classify an instrument as a sheng it must possess six characteristics: a wind chest, pipes, reeds, a blow pipe, certain tuning systems, and be made of certain materials.

Wind Chest

The sheng generally has a bowl-shaped wind chest, which can be categorized into ellipse- and circle-shaped wind chests. As mentioned above, the wind chest of the sheng excavated from the Tomb of Marquis Yi of Zeng was ellipse-shaped, but the round bowl-shaped wind chest is the most common today. Nevertheless, the gourd is considered the most representative material from which to produce a wind chest and because of this, in the Chinese musical instrument system, the sheng is classified as a gourd instrument.[36] There is a long or short blow pipe mounted on one side of the wind chest.

In the beginning of the twentieth century, folk musicians in the Henan area (located in the North China Plain) attempted to create a square-shaped wind chest (Figure 10). Local

---

[36] Since the Zhou Dynasty, Chinese musical instruments were classified into eight categories, according to the materials used in their production: gold, stone, earth, leather, silk, wood, gourd, and bamboo.

ensembles at that time often needed to play on the move like marching bands. Sometimes they would travel tens of miles and play on the move for a wedding, a funeral, or a festival ceremony.[37] Therefore, in order to help better hold the sheng and change the supporting hand during the long march, local sheng musicians worked with artisans to modify the instrument. They changed the wind chest from original round to a square shape, adjusted the original circular arrangement of sheng pipes to horizontal rows, and left a gap on both the left and right sides so players could insert their index finger. With the cooperation of thumb fingers, index fingers, middle fingers, and purlicues of both hands, the player could better hold his sheng during the performance. The innovation of a square-shape wind chest deeply influenced later designs of sheng. Not only did it introduce a principle to equally use both hands in sheng performance, but also did it break the idea that a wind chest of a sheng had to be in the original round shape. In the 1950s, Wang Huizhong (王慧中), Su Tiansheng (苏天生) and many other musicians and artisans published their new designs based on the of Henan square sheng.

---

[37] Pei Gao, "Henan Sheng Jiqi Yanzou Tese" [Characters and Performance of Henan Sheng], *Yueqi*, no. 5, (1984): 21.

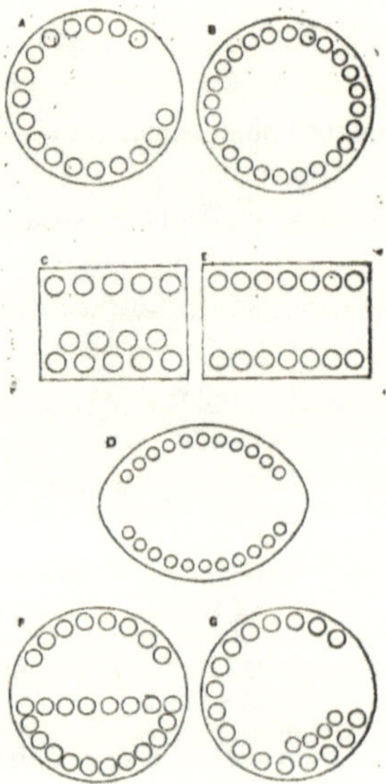

Figure 10. Vertical view of round, square and ellipse shaped wind chests.

Sheng Pipes

Multiple bamboo pipes (13, 15, 17, or more) of varying graded lengths are inserted

in an incomplete circle through the flat upper surface of the wind chest. The pipes are

arranged in two opposite triangular shapes to symbolize the folded wings of the *fenghuang*,

or the Chinese phoenix bird.[38] With a free reed secured with wax to the bottom of each

sounding pipe, the pipe on a traditional sheng sounds by the player covering a small finger

hole above the surface of the wind chest. However, some muted pipes without reeds or finger

holes remain in the design of the instrument to maintain the instrument's phoenix-like

---

[38] The *fenghuang*, or Chinese phoenix bird, are mythological birds that reign over all other birds. It is made up of the head of a golden pheasant, the body of a mandarin duck, the tail of a peacock, the legs of a crane, the mouth of a parrot, and the wings of a swallow. It is a symbol of high virtue and grace and good fortune as only appears in good time.

appearance (Figures 11 and 12).

Figure 11. Image of a Chinese Phoenix Bird.[39]

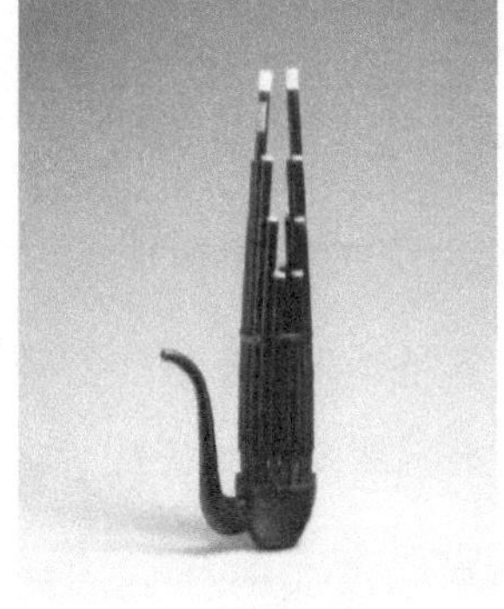

Figure 12. Sheng preserved in the Metropolitan Museum.[40]

The player can play on pitch on any of the sounding pipes of a sheng, either inhaling or exhaling. The pitch of a pipe is determined by two dependent factors. The first is an air column in the pipe, measured by length and created by an upper aperture cut through the pipe's inner wall. To create a lower pitch without making a pipe too long, the structure of folded pipe was invented. In addition to the Han Dynasty yu unearthed in Mawangdui mentioned above, the book *Sheng Fu* 笙赋 [The Ode to Sheng], written by Pan Yue (潘岳) in the Jin Dynasty (226 CE–420 CE), also mentioned about folding pipes: "脩橛内辟，馀箫外逶 [Short pipes are connected to make a 'folded' long pipe]." With the help of surviving literature, the Mawangdui yu discovered in 1980s, and modern techniques, Huang Linqian (黄林潜), Yu Xuehong (于学洪), Wang Junhua (王俊华), and many other sheng musicians

---

[39] IMGBIN, "Fenghuang Red PNG," accessed June 15, 2021,

and artisans were able to reproduce and apply such a structure to their new designs of sheng.

Sheng Reeds

The second factor that affects the pitch is the frequency of the sheng reed. The sheng reed is a rectangular tongue cut from and seated flush with the frame. It is secured with wax at the bottom of every pipe; this reed determines the vibrating pitch frequency. Due to the difficulty of changing the length of the air column or the size or thickness of the reed for tuning, a red wax dot (a mixture of wax, cinnabar, and honey) was invented no later than the Ming Dynasty (1368 CE–1644 CE). Sheng players applied the wax near the sheng's vibrating end onto the reed tongue for finer adjustments of pitch, accomplished by changing the size of the red wax dot (Figure 13). When open, the finger holes on pipes above the wind chest's flat surface function to break the air columns that reinforce the reed vibration, hence, little sound is made. In contrast, a closed finger hole seals the air column so the reed vibration is reinforced, and a clear sound can be made.

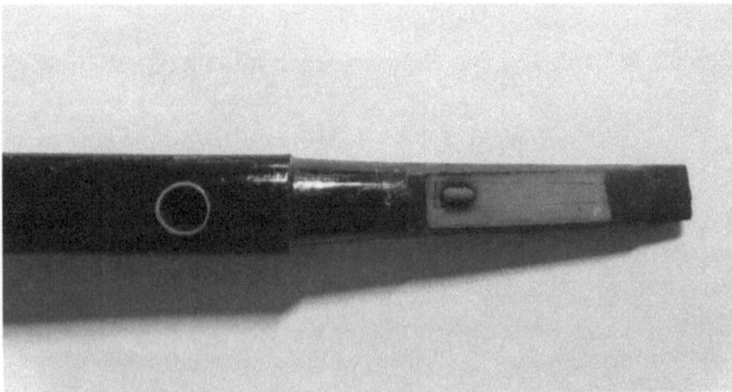

Figure 13. The red wax dot on the reed at the bottom of a sheng pipe. Photo by Haochen Qin.

The earliest reeds were probably made of bamboo. Three factors support this theory. First, the radical, or the semantic indicator of the character "簧 (reed)" is "竹 (bamboo)," which indicates a strong relationship between these two objects. A most possible explanation is that the material of a reed most likely to be used during the period this characteristic appeared was bamboo. *Shuowenjiezi*, a Chinese dictionary of the Han Dynasty confirms this theory: ". . . 簧 is the reed in a sheng. Its meaning is close to bamboo and pronounced as *huang*." The second factor supporting the theory is that the reeds on the five *yu*s discovered in the Tomb of Marquis of Yi are made of bamboo. The third factor is that the earliest clear record of copper reeds appeared in *Shengfu*: "Reeds of a sheng are made of wrought brass." Although the exact transition from bamboo-made reeds to brass-made reeds is still unknown, both extant literature and archeological discoveries confirmed that the majority of sheng reeds after the Song Dynasty were made of brass. Nowadays, due to the acoustic performance, durability, and difficulty of processing, brass is still the primary option for sheng reeds.

Blow Pipe

Until the Tang Dynasty, the sheng remained the leading instrument in court music ensembles. The performer stood or knelt while playing the sheng, holding the instrument in the hand or on the knees. Therefore, the length of the blow pipe was generally long, some could be as long as an arm (Figure 14). The Tang Dynasty sheng represented in the murals depicting court music performance preserved in museums demonstrate this point. After the

Tang Dynasty, especially in the Song Dynasty, the use of the sheng in court music began to decrease, but the musical form of small instrumental ensembles using the sheng became popular among folk people. This type of performance thus mostly accompanied folk activities, and in many cases, sheng players were even required to walk along with the crowd as they played. Thus, the length of the mouthpiece was shortened, so the player could hold the sheng and play, just as, for example, we might hold a peach with both hands for biting. In most cases, a blow pipe was made of the same material as the sheng's wind chest.

Figure 14. Image 1: Ornament on back of a bronze mirror excavated in Changdao, Shandong province. Image 2: Ensemble models made in bronze, grave good excavated in the Shaoxing, Zhejiang province. Image 3: Tomb mural from Han Dynasty in Yinan, Shandong province.[41]

Tuning System

Traditional sheng has always been tuned to the pentatonic scale. The sheng does not use a certain tone as the tonic, but adds pipes to play each pitch, using each tone of the

---

[41] Chunyi Li, *Zhongguo Shanggu Yueqi Zonglun* [A Study of Ancient Chinese Musical Instruments], (Beijing: Culture Relics Publishing House, 1996), 420.

pentatonic scale as the leading tone for different types of sheng, thus forming the sheng family: the C-tonic sheng, D-tonic sheng, E-tonic sheng, G-tonic sheng, and A-tonic sheng. If a sheng could play certain semitones, musicians would replace some pipes with "yiguan," or support pipes to help play a pitch outside the key area or even change the key area.

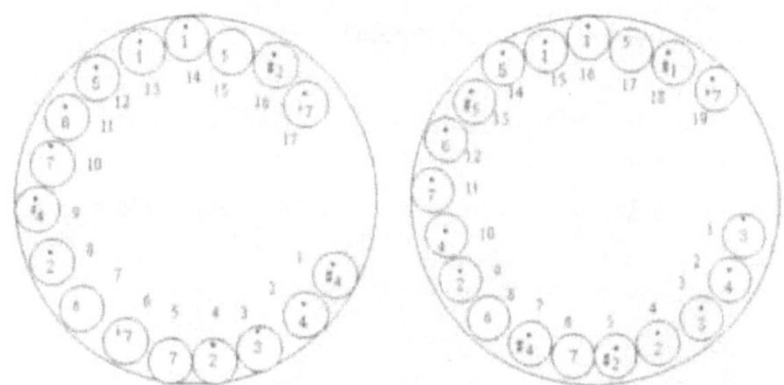

F tonic 17-reed sheng from Zhihua Temple, Beijing.1=F

F tonic 19-reed sheng in Song Dynasty, 1=F.

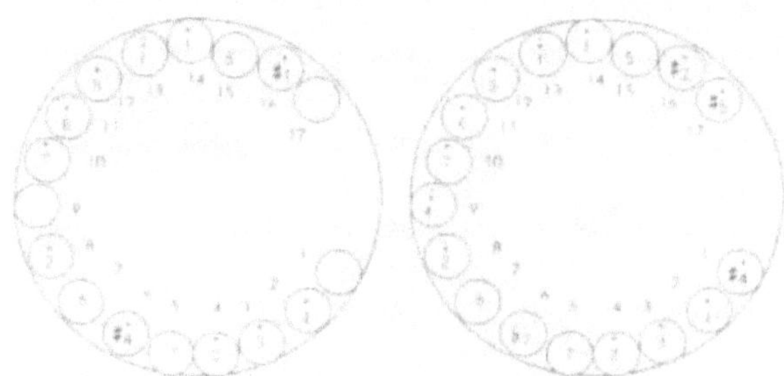

E tonic 17-reed sheng in Tang Dynasty, preserved in Shosoyin, Nara, Japan. 1=E

E tonic 17-reed sheng from Hebei province. 1=E

Figure 15. Examples of tonic sheng.[42]

---

[42] Yuejin Zhang, Lun Hebei Sheng Dui Woguo Sheng Yinwei Pailie Moshi de Chuancheng Yu Fazhan [How the Traditions of Hebei Sheng Influenced the Pitch Arrangement of Today's Sheng], *Chinese Music*, 3,

Materials

As early as in the Shang Dynasty and Zhou Dynasty, all materials needed to construct a sheng could be easily obtained from nature. The earliest material for making a wind chest was gourd. The natural hollow and spherical shape of a gourd nicely met the needs of sheng-makers, making it widely used. With the improvement of manufacturing techniques, people were dissatisfied with the irregular shapes of naturally grown gourds, so they began to fit gourds into pre-made molds before the gourd had grown, thus they would fall within accordance of the size and shape of the mold, fit to make a sheng wind chest. Later, sheng-makers began to switch to wood, lacquerware, or metal for a more regularly shaped wind chest. In this way, the materials used for sheng wind chests further developed. The sheng pipes were made of bamboo in most cases, although wood was used in rare cases. Bamboo and many kinds of metal materials are employed by sheng-makers to make the reeds nowadays, seeking to create better sounds that meet the expectations of performers.

Finally, many different parts in different textures are also required to build a traditional sheng, through a complex process; thus, the production of the traditional sheng is always difficult. Not only does it require an artisan's skills, time, and energy to build a sheng, but the instrument is also difficult to repair and maintain in good condition. Since the Tang Dynasty, China has experienced long-term, large-scale wars, so fewer and fewer artisans were capable of building a sheng; moreover, people were increasingly reluctant to learn to play the sheng because of the difficulty of its performance technique and maintenance. These factors

---

2008, 248.

combine to account for the Sheng's decline beginning after the Song Dynasty until the end of

the Qing Dynasty. The six characteristics were also starting points for the development of the

instrument and pointed the direction for later musicians in applying modifications and

innovations to the instrument.

CHAPTER III

THE EVOLUTION OF THE MODERN SHENG AFTER 1949

Reformation of National Instruments in People's Republic of China until 1953

End of the Qing Dynasty; the Last Chinese Empire (1840–1949)

The year of 1840 is a turning point for Chinese society. Before then, the majority of Chinese considered their country the ultimate power, at the center of the world, and saw little, if any, influence from foreign powers. Such a view was completely subverted after the First Opium War (1840–1842).[43] During the following 100 years, China contended with various wars, often ending as the defeated party. The series of defeats led to the Chinese identity turning from one extreme to another; that is, from the Chinese thinking of China as the leader of the world to thinking that everything *western*—products, ideas, and techniques— were better than Chinese, and that China must learn from the west. Thus, the Chinese government launched a series of reforms and modernization movements in every field, the music industry no exception.

Inspired by capitalism and nationalism ideologies, the Chinese ended their last empire, the Qing Dynasty, in 1911, and established the Republic of China in the following year. The birth of the first republic administration of China brought significant changes to Chinese society. The blind faith that a great empire would last forever was abandoned. The country and its people were no longer isolated from the rest of the world. The Provisional

---

[43] The Opium Wars were a series of military engagements fought between Britain and the Qing dynasty of China.

Constitution of the Republic of China came into effect as the temporary constitution. For the first time ever, the words "China" and "Chinese" were widely recognized by Chinese citizens, not only as regional descriptors, but also as having political and cultural implications.

Moreover, European culture and ideas flooded into the newly opened China, including European books, goods, and people. The Chinese were shocked by the strength and wealth of the western world, but gradually discovered the value of western culture as the power behind supporting material wealth, believing that western music, as a part of western culture, was superior to Chinese music. Thus, many Chinese devoted their efforts to make way for China's development through learning the best part of both Eastern and western civilizations. Musicians of the early twentieth century believed that the secret of western cultural power must lie in the differences between Chinese and western music. Instruments were considered one of the primary ways to identify and develop cultural capital. These musicians hoped to reform traditional Chinese instruments and introduce modern production methods for better performances. These ambitions initiated a movement to modify traditional Chinese instruments including the sheng in the following decades.

The Datong Music Society (大同乐社)

In the beginning of the twentieth century, unlike western musicians highly respected in China, Chinese musicians who played traditional music struggled to make a living during the upheaval of society. Compared to their colleagues playing western music, some of whom enjoyed grand halls, nice tuxedos and a good living, most musicians who played traditional Chinese music made a living by accompanying in rituals, funerals, and weddings in rural areas. Indeed, as Liu Tianhua (刘天华) wrote: "Chinese music has thousands of years of history and many excellent masterpieces; however, it has long been destroyed by continued wars."[44]

Another musician, Zheng Jinwen (郑觐文), decided to change this situation by initiating a series of modifications to traditional Chinese instruments. With friends he founded the Datong Music Society in Shanghai in 1920. He reported:

> We collected more than 120 kinds of instruments and over 1000
> excellent music pieces. I have absolute confidence that as long as we
> consult experts all around the world and examine these instruments,
> we will definitely have the ability to modify them, enhance them
> with great value, and form a beautiful, large orchestra. By then, we
> shall have a marvelous Chinese culture to show the world.[45]

By the spring of 1931, the Datong Music Society had restored 164 pieces for ancient and modern Chinese traditional musical instruments. Musicians devised a new system of musical instrument classification and categorized these instruments into nine families. A large part of such instruments were the creation and improvement of traditional musical instruments. Luo Songquan (罗松泉) and Xu Guangyi (许光毅) classified these instruments into four types: blow, plucking, bowing, and hitting. According to this classification, in total there were thirty-five plucked instruments, twenty stringed instruments, forty-three wind

---

[44] Dongsheng Pu, *Minyue Jishi 60 Nian* [The 60 Years' Chronicle of National Music] (Beijing: China Federation of Literary and Art Publishing House, 2009), 89-90.

[45] Wujia Cheng, "Ershi Niandai Minzu Qiyue Huodong Qingkuang De Huiyi" [Memoirs of Activities for Chinese Instrumental Music in the 1920s], *Journal of Nanjing Arts Institute (Music & Performance)*, no. 3 (1984): 94.

instruments, and sixty percussion instruments, all displayed in the exhibition room in Datong Music Society.[46] Due to the unstable political and financial situation, promotional efforts were limited, thus their efforts drew limited response.

Many influential associations also participated. Liu Tianhua (刘天华), Yang Zhongzi (杨仲子), Zheng Yingsun (郑颖荪), Zhao Zhiren (赵之任), Cheng Wujia (程午加), and Mei Lanfang (梅兰芳) started the Society of National Music Improvement. These musicians held weekly seminars to discuss the future of Chinese music. Cheng Wujia noted in his memoirs:

> Yang Zhongzi says: "You cannot improve Chinese music without learning western music." Zheng Yingsun says: "And you must not give up the traditional Chinese music. You must start from Guqin, Se and other Chinese instruments." Liu Tianhua says: "We need both; however, I believe what matters is we must explore something new, [something that] never happened in Chinese music." Zhao Yuanren says: "Our ultimate goal is not to study western music, but to create something good for the time. Remember, we are doing this for our country." I (Cheng Wujia) say: "Do not forget percussion instruments. These are also part of our music."[47]

Broadcasting Corporation of China's National Orchestra (中央电台音乐组国乐队)

On October 6, 1935, the Broadcasting Corporation of China set up a national music band and instrument reform team. Ten musicians were assembled to reform traditional instruments, shifting them to the twelve-tone equal temperament system. The team leader was Chen Jilue (陈济略). At its beginning only six team members formed the group, including Gan Tao (甘涛), Hu Liezhen (胡烈贞), Gao Yi (高义), Qiao Gaofu (乔告福), and Gao Ziming

---

[46] Located at No. 393 Wukang Road, Shanghai.
[47] Wujia Cheng, op.cit., 94.

(高子铭). Soon after Huang Jinpei (黄锦培), Wei Zhongle (卫仲乐), Lu Xiutang (陆修棠),

He Luting (贺绿汀), Yang Dajun (杨大钧), and Wang Peilun (王沛纶) joined the team.

In 1935, Zheng Jingwen, founder and director of the Datong Music Society, passed

away. Two years later the Japanese army captured Shanghai. The last two members of the

Society, Xu Ruhui and Zheng Yusun, left Shanghai for Chongqing, hoping to assemble a new

Society there.

In 1937, because of the Second Sino-Japanese War, members of the reform team

moved to Hankou, along with the corporation. One year later they moved again to Chongqing

and renamed the corporation the Music Group of the Broadcast Building. In 1941, the new

Datong Music Society continued in Chongqing. New members started training in classes for

amateur players and established an orphanage, hoping to train instrument-artisans. Neither

attempts were successful. The testimonies show their mission was to compete globally with

other musical cultures. In such a difficult time, Yang Yinliu wrote,

> As far as the national music is concerned, our current situation is like
> this: we have not yet fully prepared ourselves, but music of other
> cultures is already very powerful. If we don't prepare anymore, we
> can only witness the music from other cultures gradually eliminate
> the only remaining Chinese music. Therefore, we must prepare.
> However, a small amount of gradual discovery, induction, and
> development at a time seems to be far from enough to stop the trend
> of other music. Therefore, in such a situation, we should pay
> excessive attention to Chinese music. Now, it had to be taken
> seriously.[48]

The development of Chinese music remained difficult. After the Chinese victory of

---

[48] Yinliu Yang, "Guoyue Qiantu Jiqi Yanjiu" [The Future of Chinese National Music], *Musicology in China*, 4 (1989): 12.

the Second Sino-Japanese War in 1945, the Chongqing Datong Music Society partly ceased operation, and the original office of the Society in Shanghai was moved to the Li Yi National Music Research Room (礼毅国乐研究室). In 1946, along with the central government, the reform team and corporation returned to Nanjing, and Gan Tao took over as head of the music group. After some rectification, the group expanded into an orchestra, but soon disbanded at the end of 1948 due to insufficient financial support.

In the early twentieth century, although musicians were enthusiastic on the topic of improving traditional Chinese musical instruments, most areas of China had been at war for more than two decades. Chinese musicians remained passionate about working in any conditions, but issues as complex as musical instrument modification were inevitably put on hold. No evidence has appeared to demonstrate modifications to the sheng during this period; however, the preparation for modifications had occurred. Although the exact direction of instrument reform had yet to be determined, the necessity was already agreed on by a consensus of Chinese musicians. Nevertheless, organizations mentioned above trained a large group of talented practitioners who later became composers, players, and instrument-artisans, and contributed extensively to the reform of traditional Chinese instruments.

People's of Republic of China and Influence of the Soviet Union (1949–1953)

After years of domestic war against the administration of the Republic of China, the People's Republic of China was founded in 1949. Most of China was again peaceful, thus the public's demand for musical instruments greatly increased. The traditional instrument workshops in Beijing, Shanghai, Guangzhou, Tianjin, and other places accepted a large number of government orders. Even workers in the metal processing, and gold and silver jewelry shops, came to participate. The formerly scattered and declining musical instrument production began to take shape.

The members of the former Datong Music Society and other traditional music associations, especially the skilled workers who worked in instrument manufacturing, entered traditional instrument factories, passed on their skills to apprentices, and participated directly or indirectly in important music reform projects. As traditional instrument production increased, cities like Yingkou, Tianjin, and Guangzhou became centers of production.

In October, 1952, four groups of artists from the Soviet Union came to China, participating in the China-Soviet Union Friendship Week. During their stay, Soviet music authority Mikhail Chulaki Ivanovich and Boris Zakharoff gave lectures to Chinese musicians on the experience of reforming folk music, instruments, and bands in the Soviet Union and Azerbaijan.

Considering the elements of musical cohesion, particularly that when the timbre of a group of instruments is relatively uniform the harmonies sound much better, Russian musicians had created three kinds of *domra* and five kinds of *balalaika* to form a large Russian ensemble, together with a keyboard, accordions, and percussion instruments. This new type of ensemble was capable of performing traditional folkloric music, as well as Tchaikovsky's Symphony No. 4. When speaking of their experience reforming traditional instruments, Chulaki said:

> There are seventeen pitches within an octave when you listen to the
> authentic Azerbaijan folk music. This makes it impossible for
> musicians to play such instruments in twelve-tone equal
> temperaments. So, they decided to reform their own music
> according to the twelve-tone equal temperament. Now the gap
> between the Azerbaijanis and other people is gone, so they can play
> more advanced instrument to tell their story.[49]

Meanwhile, similar messages from some neighboring Asian countries were also

spreading throughout China. In Kazakhstan, musicians applied different sizes of domra to

form a distinctive national band. In Indonesia, the folk musicians in Jakarta reformed a full

set of more than thirty angklungs to use twelve-tone equal temperament, forming a range of

nearly three octaves, which made possible not only highlighting the main melody, but also

playing complex harmonies. In Japan, Noru Mitsuki remembers the efforts from his fellow

musicians:

> In Japan, in 1964, we formed the Japan Music Group . . . for over 30
> years, we have roughly completed the techniques of modern
> ensemble with Japanese instruments. Japanese instruments have the
> techniques and many works that can cooperate with various western
> instruments and even large symphony orchestras.[50]

If the introduction of western technology and music theory made the reform

possible, then the message of other Asian countries in progress, or even completing their

instrument reforms, put more pressure on China. China's instrument reform became not

simply about the technical practice of musical instrument performance, but also a symbol of

---

[49] Yuanqing Li, "Tan Yueqi De Gailiang Wenti" [On the Modification of Traditional Instruments], *People's Music* no. 1 (1954): 25.

[50] Noru Mitsuki, Yanqiao Wang, Lin Gong (Trans.), *Japan Organology: Preface* (Beijing: People's Music Publishing House, 2000).

social modernization and cultural modernization.

Subsequently, such music reforms began to expand gradually, and the Soviet music reform group's knowledge and experience remained highly valued. In 1954, the Music Reform Group was established in China, and the first Musical Instrument Improvement Conference held. Vice Minister Ding Xilin (丁西林), of the Ministry of Culture, proposed that instrument reform must focus on the needs of performance and preservation, and carry forward the characteristics of the culture while also carefully considering the experience of the Soviet group.

By 1953, the All-China Music Workers Association was reorganized into the Chinese Musicians Association. Among its six planned tasks, the reform of musical instruments was listed as fifth. Large-scale music reform was on the agenda in terms of organizing members to promote the study and sorting of classical music, song and dance music, opera music and other folk music of various ethnic groups in the country, and to improve traditional instruments. Thus the Association sought to carry forward the excellent traditions of various ethnic music in the country.

Beginnings of Instrument Reform (1954–1966)

In 1954, Chinese practitioners, began discussing instrument reform and emphasizing its importance at the symposium of the National Traditional Opera, Folk Song and Dance Performance.[51] The National Music Association and the Central Conservatory also

---

[51] The symposium happened in October 1954, after the Second Congress of the China Federation of Literary and Art Circles from September 23rd to October 6th in 1954 in Beijing.

began receiving suggestions and designs on the improvement of instruments. The reform of musical instruments became a customary topic needing urgently to be resolved in the music industry during this period. Thus, the central government decided to set up special research institutions to serve and promote reforms of instruments.

In October 1954, at the suggestion of Li Yuanqing (李元庆), the National Music Committee of the Chinese Musicians Association, along with the National Music Research Institute, held the first national symposium on the improvement of national musical instruments in Beijing. During the four-day meeting, more than seventy reformed musical instruments (including finished products and design drawings) were introduced and auditioned. Over fifty people from related fields, including instrument-artisans, practitioners, instrument technicians, salespeople, and composers participated in a series of meetings. This was the first time in Chinese history that an event was held specifically for the improvement of musical instruments.

Li Yuanqing published "On the Musical Instrument Reform" and "The Principles of Instrument Reformation," providing theoretical support for the ideas and principles of musical instrument reform according to suggestions given by Soviet experts. Objectively speaking, these policies form a summary of the experience of instrument reform in the Soviet Union, especially the idea of building an entire family of musical instruments based on a certain musical instrument as a prototype. These policies might be regarded as the work of Chinese practitioners learning from the west and imitating the experience of the Soviet Union in reforming folk instruments.

Inspired by the national movement of traditional instrument reforms, Hu Tianquan (胡天泉), a sheng player from Shanxi, in 1956, sought to compose a sheng solo, entitled "The Phoenix Unfolds" 《凤凰展翅》. However, even as a professional sheng player, the only thing he had at hand was one of the most traditional and common sheng in the 1950s, with seventeen pipes and thirteen reeds. Hu Tianquan remembers the difficulties he was facing:

> I found the instrument I have was limited in volume and range. Some reeds could not vibrate steadily when I was playing. I tried to attach reeds to those muted pipes respectively, thinking at least I will make every pipe I have sounds.[52]

In truth, most traditional sheng used in 1960s were more or less like this one: a small sheng with seventeen pipes, four of them muted; the thirteen playable pitches fell into the same single octave, and there were repeated pitches among the thirteen pipes. There were no half steps so the composers and players could not modulate when playing, severely limiting the use of sheng, and creating a stereotype of the sheng as good for accompanying only certain instruments in certain pieces.

In composing "The Phoenix Unfolds," Hu Tianquan consulted several types of sheng fingerings, including the Shanxi, Hebei, Shandong, Northeast sheng, and the Henan square sheng, as well as the records of the Tang dynasty sheng. After serious analysis and more research, he determined the most suitable fingering arrangement for his model of sheng that would work best for his composition.

In order to maintain the original beautiful appearance of the sheng, he installed all

---

[52] Shengnan Ma, "The Adventurer of Chinese Sheng" (Master's thesis, Shandong Normal University, 2011), 47.

the keys (or buttons; later referred to a type of sheng that is "keyed") on the inside of the

pipes, to keep the sheng as small, lightweight, and beautiful as possible. In order to make

uniform the sound and balance the volume, with the help from Ni Zhishan (倪志珊), Hu

Tianquan added a metal tube (called a *kuoyin guan*, or resonating tubes) to the treble area,

installed opposing "resonance tubes" in the midrange and bass areas, and opened up the two

adjacent pipe walls so that the two pipes aided each other in resonating. The sheng reformed

in this way had a mellower and purer sound quality than sheng before it. With the use of

opposing resonance hoods and resonance through pipes in the middle and bass areas, the

volume of each sound area tended to be more balanced. With the ideal to avoid disrupting the

appearance or timbre of the instrument, Hu made bold changes, broadened the range of the

instrument, and enhanced its function.

The Cultural Revolution (1966–1976)

During the Cultural Revolution, from 1966 to 1976, under the policy of the Gang of

Four that everyone "stop work and carry out revolutionary activities," most of the instrument

production in mainland China came to a standstill.[53] This stagnation in the musical

instrument-related industries grew even more serious than in other industries.

In the discourse propagated by the Gang of Four, western musical instruments

became a symbol of "corrupt capitalism," and traditional Chinese musical instruments a

---

[53] Central Cultural Revolution Small Group, Yongyi Song ed., *Guanyu Gongchang Wenhua Da Geming De Shier Tiao Zhishi* [Twelve directives regarding the Cultural Revolution in factories (draft)], Chinese Cultural Revolution Database,

symbol of the imperial culture of old China. In other words, during this decade most of the musical instruments that people might come into contact with were defined as cultural symbols that represented "degenerate thoughts" and should be resolutely eliminated for New China ideals. Even the use of musical instruments was strongly prohibited, and experts engaged in musical instrument production and technical research were also criticized. In this social context, relevant practitioners were forced to deprive themselves, staying away from work to ensure their own safety. A large number of instrument manufacturers were forced to suspend production, many performance groups forced to disband, and the few surviving groups and individuals almost completely stopped music-related activities. Many professional musicians were sent to factories and rural areas for "ideological reform."[54] Sheng instrument reform temporarily went dark.

However, with the development of the Cultural Revolution, the Gang of Four gradually changed their attitude to traditional Chinese music and instruments. The suppression of the development of traditional music was thus eased.[55] Even if the Gang of Four had failed to recognize the value of traditional music, they admitted that the genre was the most familiar to ordinary Chinese people. And they needed this widely recognized art form to create and perform works that conformed to their political thought, to deliver political propaganda to ordinary people. Thus, even as the Gang of Four released some

---

[54] Shu Gao, "National Musical Instrument Reform—the Chronicle of Putting," PhD diss., (China Art Academy, 2012), 22.

[55] Ming-yen Lee, "Political Influence on Music Performance: The Development of Modern Chinese Orchestra in Mao's Regime (1949-1976)," *Malaysian Journal of Humanities and Social Sciences*, Vol. 6, No. 2 (2017): 7.

restrictions on traditional music in 1972 or allowed all parts of the country to organize their own literary and artistic teams to distribute political propaganda in traditional instruments, folk music finally won a living space for itself.

This situation brought negative and positive effects on the development of traditional instruments. On the one hand, the involvement of political forces restricted the direction of the development of instruments and forced the reform of these to serve political purposes. To be precise, the development of "national" musical instruments was limited to the formation of specific bands for specific repertoire and specific types of performances. Even the reform work was limited to specific purposes, especially the requirement to transform as many as possible traditional musical instruments into a bass version to fill gaps in bass parts of the national orchestra ensemble. This reform greatly restricted the attempts of musicians to freely explore various developmental possibilities of traditional instruments, and directly led to fewer reforms of these during this period.

On the other hand, the formation of a large number of ensembles called for mass production of various instruments. Thus the production of traditional instruments improved efficiency, and also resulted in pressure for publication of industry standards. It also resulted in focusing instrument reform, beginning to shift from the function of the instrument itself to the mechanization and standardization of instrument production. The stability of the structural performance of the musical instrument itself, and the difficulty of use and promotion eventually becoming central to industry discussion. Still, considerable progress was made in the various machines of production, processes, and industry standards.

The development of the sheng fell in line with this trend. In terms of specific musical instruments, only the fifty-two-tone keyboard, or rehearsal sheng, an improved sheng created for playing "model operas," appeared in the entire ten years of reform. But at the same time, the manufacturing process related to the sheng made significant progress. Sheng-makers represented by Sun Rugui created machinery, such as angle cutters, which greatly improved the production efficiency of the sheng. In addition, after long-term thinking and accumulation, sheng musicians and artisans initially formed their own plans for the development of the instrument, and only the end of the Cultural Revolution offered them opportunities to fulfill their ideas.

The Rapid Development of the Sheng after 1976

After the Cultural Revolution, the entire music industry entered a stage of recovery and development. According to the No. 76-6 file, published by the Ministry of Light Industry in People's Republic of China, since 1976, the Beijing Musical Instrument Research Institute became a national gathering place for scientific and technological information on musical instruments; in fact, it became the main organization for musical instrument reform. In December 1978, the Acoustic Society of China, the Central Conservatory of Music, the Music and Dance Office of the Institute of Literature and Art of the Ministry of Culture (now the Institute of Music of the China Academy of Art), and the Beijing Musical Instrument Research Institute gathered in Beijing to establish the Beijing Musical Instrument Association. The main work of the association was to participate in the development of materials for making musical instruments and the formulation of standards for the instrument industry. This cooperative and participatory working method differed from the working method of the National Music Research Institute in the early days of the People's Republic. The former failed to lead the entire reform activities as the latter did to participate equally in various market forces, such as research institutes and instrument production plants. The post-mortem evaluations and ratification of the existing reform results were also included in the work of the association.

At the end of 1978, the Third Plenary Session of the 11th Central Committee of the Communist Party of China was held. The social order had gradually returned to normal, and the production of musical instruments kept increasing, but the subsequent implementation of the reform and opening policy also greatly increased the domestic market's demand for western musical instruments, which then reduced the development space for traditional musical instruments. Many traditional instrument factories had to switch to producing western musical instruments in order to survive. In 1980 alone, more than 100 traditional musical instrument manufacturers across the country turned to the production of pianos and related accessories. The survival and development of national music instruments again faced tremendous pressure. However, unlike the overwhelming and political pressure faced during the Cultural Revolution, this time these instruments struggled to compete with their western "peers." From the perspective of market competition, this pressure was also the biggest driving force for the development, reform, and progress of traditional musical instruments.

Under such a social background, since 1976, traditional Chinese musical instruments, including the sheng, have undergone a period of rapid development. This rapid development of the sheng, lasting over five decades now, resulted from the following factors:

1. Accumulation of experience in mass production technology

As mentioned above, for the purpose of using traditional music for political propaganda as early as the Cultural Revolution, political leaders asked the producers of traditional musical instruments to increase production efficiency and to produce in numbers and in best quality traditional musical instruments for propaganda personnel use. Since then, sheng-makers sought to standardize large-scale instrument production. By the end of the Cultural Revolution, the beginning of reform and opening in the 1980s, the sheng artisans, from both instrument factories and private workshops had accumulated a considerable degree of experience and had also invented some production machinery.

2. Music reform information-exchange seminars

The activity of so many music reform information-exchange seminars allowed the participants of sheng reform to exchange ideas with their peers, sparking invention and change. Since 1976, various conferences on the musical instruments reform, including the National Musical Instrument Industry Science and Technology Information Work Conference, the Beijing Musical Instrument Society Annual Conference, the National Minority Art Performance Conference, and the National Minority Musical Instruments Introduction and Exchange Conference, have proliferated, and reform content involves acoustics and history, classification, performance research, archeology, production and management and other aspects of these instruments.

As the core unit of this reform work, the Beijing Musical Instrument Society held symposiums on such reforms almost every year between 1977 and 1984. At the meeting, experts across the country exchanged opinions, confirmed each other's ideas, and produced many far-reaching results. The National Music Committee of the Chinese Musicians Association, the Science and Technology Bureau of the Ministry of Culture, and the Beijing Musical Instrument Society jointly organized a national symposium on the reform of traditional wind instruments in Suzhou, held September 20–26, 1981. After the symposium, these parties formed a joint design team for *dizi* and sheng in Shanghai. In September 1981, Zhou Weizhi (周巍峙) put forward four principles for the improvement of traditional musical instruments in the summary report of the Jinan National Youth Folk Musical Instrument Solo Performance Observation Meeting. These included preserving the characteristics of timbre, maintaining the basic form, retaining the performance skills, and considering needs of both professional and non-professional players.

3. Establishment of award system for scientific and technological inventions

Although the reform of these instruments was in full swing in the academic world, the difficult situation faced by factories had not improved because the reform of musical instruments had not produced mature products for a period. At the end of the 1970s, there were fewer than twenty of these instrument factories in China, and the surviving factories were in poor financial condition, no longer able to invest funds in research on instrument reform. Therefore, the government set up bonuses for the collection and sorting of these instruments, reforms, new inventions, and other research purposes.

4. National testing and recognition systems gradually maturing.

In 1978, the official document of the National Science and Technology Commission emphasized that all scientific and technological research results, including the reform of musical instruments, would be divided into four levels: national, provincial, municipal, and general, according to the importance of the project and the size of the area involved. The process involved attaching the relevant report, calling the corresponding level of professionals for technical appraisal, registration and filing, and awards.

In November 1979, by invitation of the Beijing Musical Instrument Society, and hosted by the Central Conservatory of Music and Beijing National Musical Instrument Factory, the first large-scale appraisal meeting of professionals from the central, local, military, art academies, and art groups was held in Beijing. This was a successful event. At the appraisal meeting, each musical instrument was accompanied by a design drawing; the inventor, the testing unit, and the assisting unit were announced, and various performance and parameter changes before and after the test were recorded for experts to review.

Since then, the musical instrument appraisal meeting has become a widely-recognized protocol –not officially required though—after the musical instrument reform. Until today, the achievements of musical instrument transformation from various places are reported to special institutions in Beijing for registration and filing, and experts from music schools, musical instrument factories, musician associations, professional orchestras and other fields are organized to conduct public appraisal and review.

Since then, the government's role in the reform of traditional musical instruments has officially changed from leading to supervising and establishing standards. The first-line

sheng users and producers from scientific research institutes, professional academies,

performing arts groups, musical instrument factories, and even individual workshops, have

emerged in large numbers to show their hard work over the years.

New Morphologies of the Sheng since 1949

Twenty-six-Reed Round Sheng (圆笙) | Zhao Dezhen (赵德震)

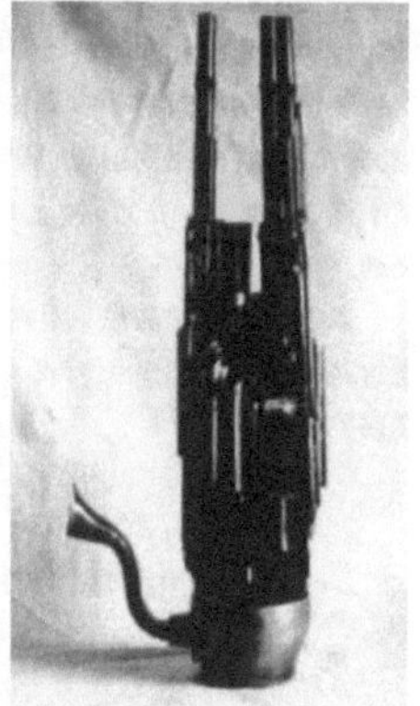

Figure 16. Zhao Dezhen's Twenty-six-Reed Sheng.[56]

Around 1956, Zhao Dezhen, of the Liaoning Song and Dance Troupe, developed

the twenty-six-reed round sheng with extra buttons. His design broke with the pitche

arrangement of the traditional sheng; it is arranged symmetrically from the middle (bass) to

both sides in twelve half steps, and has a range of C1–D3, over two octaves, including all

(complete) half steps. In 1953, Zhang Zirui (张子锐) used the ancient rule of Six Lǜ (六律)

---

[56] Guangliu Li, *Zhongguo Sheng Yishu* [The Art of the Sheng], (Beijing: Culture and Art Publishing House, 2006), 462.

to arrange the keys on a big dulcimer. Zhao Dezhen's pitch arrangement was perhaps inspired by Zhang Zirui's dulcimer. This sheng did not become widely known, but later sheng-artisans inherited the idea of the two-sided symmetrical pitch arrangement.

Holding Sheng (抱笙) and Lined Sheng (排笙)

The holding sheng instrument family includes the thirty-six-reed alto sheng, the thirty-two-reed bass sheng, and the twenty-four-reed bass sheng. Zhang Zirui (张子锐), employed by the Central Broadcasting National Guan Yanqiao Orchestra, was inspired by the Miao people's (苗族) *lusheng* (芦笙) band, with a total range of more than five octaves, covering high, alto, and bass ranges. In 1955, the resonant tube (amplifier tube) of the lusheng was transplanted to the traditional sheng. In the following year, Wu Zhongfu (吴仲孚) and Sun Rugui (孙汝桂), from the Beijing National Musical Instrument Factory, combined the amplifier tubes from the lusheng band instruments (including the treble and bass lusheng) with the traditional sheng, thus the *bao sheng* [holding sheng] and *paisheng* [lined sheng] were developed. In 1961, Zhang Zirui was transferred to the Suzhou National Musical Instrument Factory, and Yan Genshan (严根山) and Qiang Rukang (强汝康) were put in charge of sheng reform.

After the holding sheng and lined sheng were made, Wang Junting (王俊亭) of the Shanghai National Orchestra was first to play them on the stage. The sheng bodies were made up of sheng pipes, reeds, amplifier tubes, wind chest, and buttons. In addition, there was a mouthpiece, and the buttons used mechanical devices to control opening and closing the valves to produce sound.

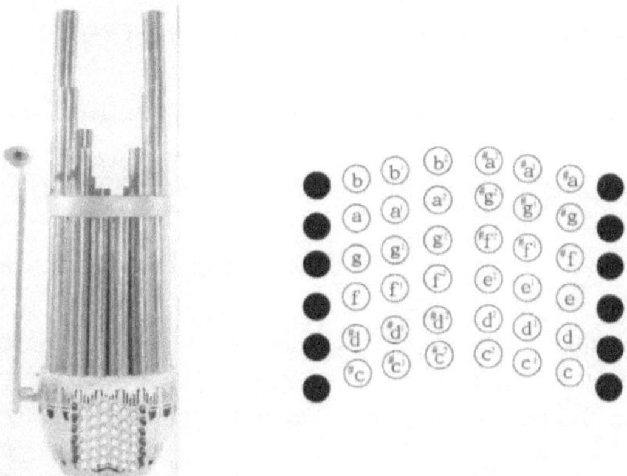

Figure 17. Holding sheng and its pitch arrangement.[57]

In the arrangement of pitches, the ancient method of Six Lǜ (六律) was used to

divide the pitches into two symmetrical groups: "Six Lǜ" and "Six Lǚ," a half-step

difference, and the adjacent keys of the same group a major second difference. Three keys in

a different octave are connected in three ways (in parallel, horizontally, and vertically). Those

on the lined sheng are placed horizontally, and those for the holding sheng are placed

vertically; thus the stacked row fell out of use.

In terms of shape, the designers adopted the body of the round sheng for the

holding sheng, and used a symmetrical arrangement of panpipes for the lined sheng. The

shapes of these two kinds of sheng are different, but the acoustic principle and internal

structure are exactly the same, as are the performance factors and sound. The two kinds of

sheng can play four voice types: bass, tenor, alto, and treble. The bass range is D1–D, the

---

[57] Dongsheng Liu, *Zhongguo Yueqi Tujian* [Chinese Musical Instruments Illustration] (Qingdao: Shandong Education Press, 1992), 156.

tenor C–D1, the alto C–B2, and the treble G–F#3, all with complete octave half steps.

After Zhang Zirui left from the Central Broadcasting Group, Wang Zhongbing (王仲丙), Yang Jingming (杨竞明), and Wang Linan (王力南) took over the reform work of the sheng. They continued to cooperate with the Beijing National Musical Instrument Factory, and later completed the Thirty-six-Reed Alto Lined Sheng. For this instrument, each pipe is equipped with a reed and a copper amplifier tube, which is inserted in the flat square wind chest. The pitches are also arranged in a way as to divide the keys into three parallel rows on both left and right sides; each group has six keys for a total of thirty-six keys. The keys are in a vertical format, the range covering nearly three octaves, from G to F#2 with all half steps available. This kind of sheng was used by many Chinese orchestras in 1960s to 1980s, but because the vertically arranged keyboard was not as convenient as the horizontally arranged keyboard, it was gradually eliminated after the 1980s.

Among those produced by the Suzhou National Musical Instrument Factory, the Thirty-six-Reed Alto Holding Sheng, with a range of D–C#3 most represents the outset of this production. By 1977, the factory developed the Thirty-two-Reed Bass Holding Sheng, whose body amplified the Alto Holding Sheng. The Bass Holding Sheng has sixteen pitches on both left and right sides, divided into three rows of white keys and one row of black keys. The black keys interlock, thus by pressing a black key the player can simultaneously sound three tones that range two octaves. The entire range includes C–G1, with complete half steps covering up to two octaves. These two kinds of holding sheng are still used in many Chinese orchestras, and among all the lined sheng produced by this factory, the Thirty-six-Reed Alto lined sheng is the most popular.

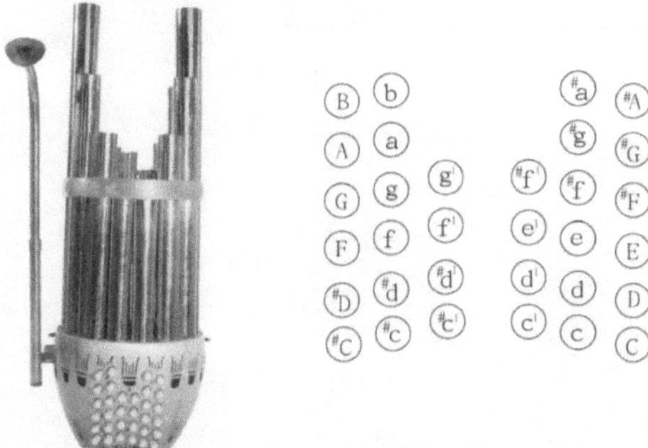

Figure 18. Bass holding sheng and pitch-arrangement.[58]

In 1996, Chen Huanhui of Taiwan Xianjin Musical Instrument Company combined

the structure of Thirty-six-Reed Alto Sheng with the range of Thirty-two-Reed Bass Sheng to

form the Thirty-two-Reed Bass Sheng. This design is still manufactured by Suzhou National

Musical Instrument Factory, with a range of C–G1, covering up to two octaves with complete

half steps. The key system is the same as the Thirty-six-Reed Tenor Sheng.

The new lined sheng's range and structure are the same as the alto lined sheng, but

its mechanical structure is quite different. In the design of the button arrangement, the scale

goes from left to right, and from front to rear alternately ascending: the first row is the bass,

the second row is midrange, and the third row is treble. The buttons lie between the player

and the amplifier tube, to allow the player to see the keys clearly and limit the performance

---

[58] Dongsheng Liu, *Zhongguo Yueqi Tujian* [Chinese Musical Instruments Illustration] (Qingdao: Shandong Education Press, 1992), 156.

difficulties in practice. Therefore, this design is widely accepted in conservatories and schools, although the sound is not as strong as that of the Holding sheng.

Alto Holding sheng and Lined Sheng

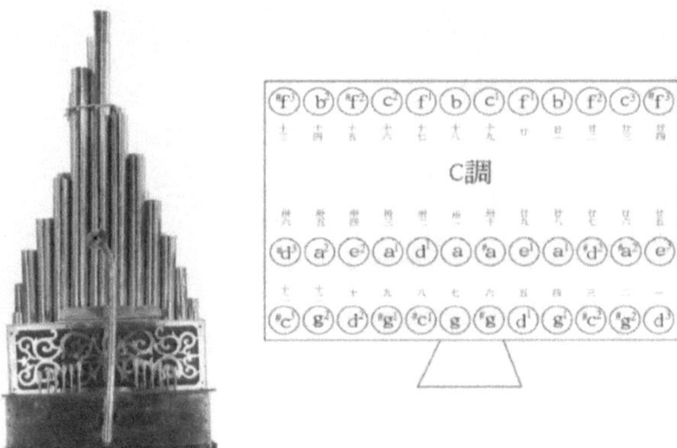

Figure 19. Alto Lined Sheng and Pitch-arrangement.[59]

Wang Linan (王力南), Wang Zhongbing (王仲丙), and Yang Jingming (杨竞明) from the Central Broadcasting National Orchestra cooperated with Beijing Musical Instrument Factory to develop a thirty-six-reed alto lined sheng. Each pipe has a copper-made reed and resonating tube. All thirty-six pipes are vertically installed into a flat rectangular wind chest. The pitch-arrangement is the same as the lined sheng mentioned above, and the softer sound is easy to integrate with other musical instruments. However, as a mid-bass instrument, its volume is too low and the vertically-installed keys are not easy to perform,

---

[59] Dongsheng Liu, *Zhongguo Yueqi Tujian* [Chinese Musical Instruments Illustration] (Qingdao: Shandong Education Press, 1992), 156.

which made it abandoned by orchestras in the 1980s.

Twenty-four-Reed Round Sheng | Li Ying (李英)

In 1958, based on the design of the square and round sheng, Li Ying used the square sheng's three-row pipe arrangement for the design of the round sheng, in accordance with the twelve-tone equal temperament, and developed the Twenty-four-Pipe Round Sheng. There are nine pipes in both the front and rear rows, and six in the middle row. Both hands can reach the pitches from the left or right end of the instrument to play the correct pitch. The range covers D1 to C#3, nearly two octaves with complete half steps. This sheng was used by the Chinese Song and Dance Theater from the 1950s to the 1970s, but it is now replaced by other kinds of sheng.

Turntable Keyed Sheng | Sun Binglin (孙炳麟) and Sun Rugui (孙汝桂)

In 1959, Sun Binglin of Beijing No. 1 Musical Instrument Production Cooperative, and Sun Rugui of Beijing National Musical Instrument Factory, jointly developed the Turntable Keyed Sheng. This sheng's copper wind chest was divided into two parts that could rotate freely. The upper half of the wind chest is marked with keys. The player presses the single key to play a single-tone melody, and the co-movement key to play traditional harmonies in fourth, fifth, and octave chords. By turning the upper part, all twelve keys can be played when a single fingering is mastered. When the key is modulated, the upper wind chest must be turned to match the key before the sheng can be played. Although this sheng can play all keys, it is impossible to practice rapid or frequent modulation, nor can it play any harmonies or chords freely, thus, it was never widely used.

Keyboard Sheng | Shenyang Conservatory of Music

In 1959, the Shenyang Conservatory of Music created an organ-style "keyboard sheng" made from parts of an organ. The pipes are placed on the back of the organ's frame and the organ reeds are replaced with sheng reeds. The pedal is used to manipulate the bellows to produce sound. The range is four to five octaves with complete half steps. Changing the sheng this way, to such a keyboard-controlled instrument, resulted in lost characteristics of the original instrument. Moreover, the performance techniques of the sheng, including fingering, performance posture, and performance skills, were also changed. Therefore, this model was never popular or accepted by performers.

Twenty-four-Reed Square Sheng in D | Wang Huizhong (王慧中)

In 1963, Wang Huizhong developed the Twenty-four Reed Square Sheng in D, on the basis of the fifteen reeds sheng in the key of D. The range lies from D1–F#3, reaching more than two octaves, but with incomplete half steps and difficulty in modulation. At first Huizhong used it himself with the Central Chinese Orchestra, but now it has been replaced by a newer sheng design that has a better performance sound.

Thirty-two-Reed Plus Key Square Sheng | Yang Daming (杨大明)

Yang Daming, a member of the Central Broadcasting Nationality Management

Group, developed the Twenty-five-Reed Sheng in the 1950s. In 1963, he designed the Thirty-two-Reed keyed sheng with a square wind chest. The designer adopted Zhao Dezhen's (赵德震) symmetrical pitch arrangement, the range is D–D3, nearly three octaves with complete half steps. This sheng was once used by players in the Beijing Central National Orchestra and was quite popular, but it was replaced by a newer style sheng with better performance.

Twenty-four-Reed, Keyed Amplified Sheng, Thirty-six-Reed, Keyed Amplified Round Sheng, and Resonating Tubes (Amplifier Tubes) | *Zhang Zhiliang (张之良) and Sun Rugui (孙汝桂)*

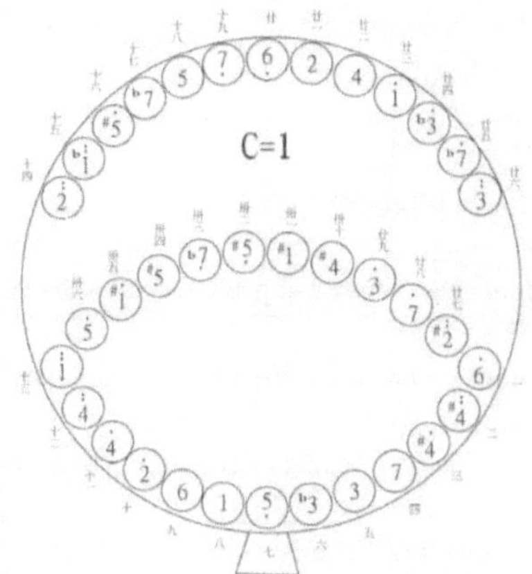

Figure 20. Thirty-six-Reed Keyed Amplified Round Sheng.[60]

Since 1970, Zhang Zhiliang of the Central Conservatory of Music and Sun Rugui of the Beijing National Musical Instrument Factory successively developed the Twenty-four-Reed, Keyed Amplified Sheng and the Thirty-six-Reed, Keyed Amplified Yuan Sheng. The

---

[60] Dongsheng Liu, *Zhongguo Yueqi Tujian* [Chinese Musical Instruments Illustration] (Qingdao: Shandong Education Press, 1992), 156.

Thirty-six-Reed, Keyed Amplified Round Sheng design is based on the pitch-arrangement of the Hebei round sheng and the Henan square sheng. The range is extended to three octaves, thus G–A3. This model uses keys and an amplifier tube is installed on the sheng pipe. The short blow pipe was adopted for the design and is supported on the thigh with a frame.

The Thirty-six-Reed, Keyed Amplified Round Sheng was also popular in its early years, but later replaced by a newer sheng with better performance. For playing traditional repertoire, many players prefer the Twenty-four-Reed Keyed Amplified Sheng. As for the amplifying tubes, the cylindrical, open-end amplifying tube provides the largest volume and a thicker tone, so was later generally adopted for all sheng. For the Thirty-six-Reed, Amplified Round Sheng, the designers adopted the wind chest of the round sheng of Hebei, and on the basis of its traditional pitch-arrangement, increased the number of pipes to thirty-six. Six different shapes of amplifying tubes have been tested: cylindrical wrapping around the pipe, open-end cylindrical, short cylindrical, close-end shallow bowl, open-end shallow bowl, and the plastic shallow bowl. The bowl made of celluloid makes the sweetest and softest sounds, with a moderate volume that best fits other instruments in an orchestra.

Thirty-two-Pipe with Thirty-four-Reed, Keyed Amplified Sheng | Fang Pudong (方浦东) and Sun Rugui (孙汝桂)

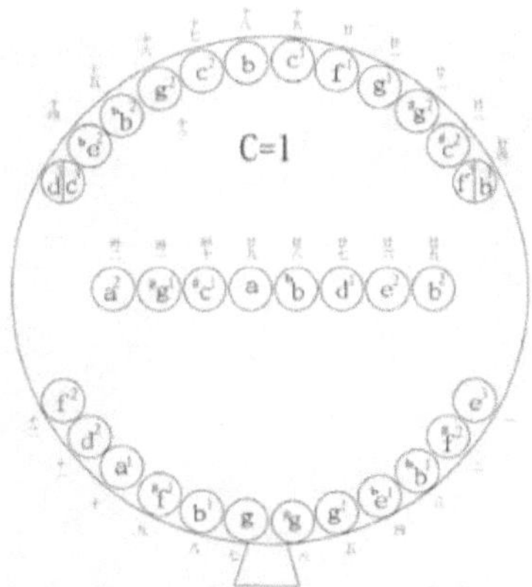

Figure 21. Thirty-four-Reed Sheng.[61]

From 1972 to 1975, Fang Pudong and Sun Rugui successively developed the Twenty-

five-Reed Keyed Amplifying Sheng and the Thirty-two-Pipe and Thirty-four-Reed Sheng,

based on the design of Li Ying's Twenty-four-Reed Sheng. The pipes of a thirty-two-pipe

sheng are divided into three rows: the front and rear rows have twelve pipes arranged

separately, and the middle row has eight pipes. The first and last pipe of the rear row have

two reeds within them. Seven linkage keys were added to solve the problem of unplayable

chords.

The range is G–F3, nearly three octaves, from G to C#3, with complete half steps.

The Thirty-two-Pipe, Thirty-four-Reed, Keyed Amplified Sheng has been used by the

Chinese National Orchestra of the Chinese Opera and Dance Theater, and the Hong Kong

Chinese Orchestra since the 1970s, later replaced by a newer style sheng with better

performance.

---

[61] Dongsheng Liu, *Zhongguo Yueqi Tujian* [Chinese Musical Instruments Illustration] (Qingdao: Shandong Education Press, 1992), 156.

Family of Thirty-six-Reed, Keyed Amplified Square Sheng | Sun Rugui (孙汝桂), Wang Huizhong (王慧中), Yang Daming (杨大明)

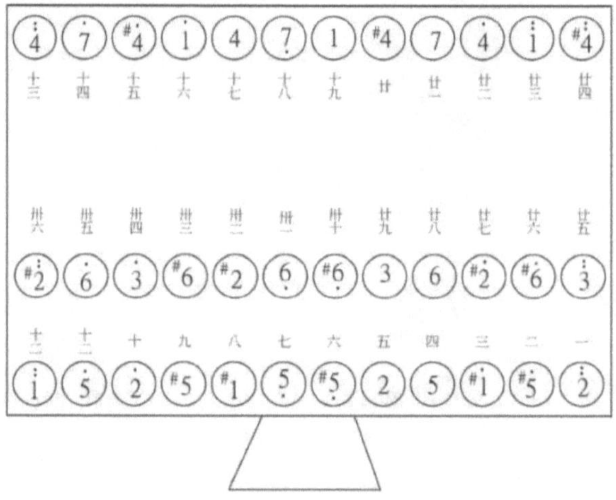

Figure 22. Thirty-six-Reed Keyed Amplified Square Sheng.[62]

In 1972, Wang Huizhong, of the Central Chinese Orchestra, redesigned the Thirty-six-Reed, Keyed Square Sheng, based on Yang Daming's Thirty-six-Reed, Keyed Square Sheng. In 1974, Sun Rugui of Beijing National Musical Instrument Factory completed the design of the amplifier tube and the production of this sheng. This sheng is composed of wind chest, pipes, hoop parts, frame, key, pipe style amplifying tubes, blow pipe, and other parts. Using the shape of a square sheng, the thirty-six pipes are divided into three rows, front, middle, and rear, arranged in parallel, and each pipe is equipped with a levered key and an amplifier tube. The pitches are also arranged in Zhao Dezhen's symmetrical arrangement.

---

[62] Dongsheng Liu, *Zhongguo Yueqi Tujian* [Chinese Musical Instruments Illustration] (Qingdao: Shandong Education Press, 1992), 156.

The twelve half steps are arranged symmetrically from the middle of the wind chest to the outside, and the adjacent keys are a major second apart. The sound range is G–F#3, nearly three octaves with complete half steps.

After its appearance, this kind of amplified sheng gradually replaced other similar treble, keyed sheng, and is now widely used by various orchestras and performers. In 2000, Sun Rugui, Wang Huizhong, and Yang Daming together completed the development of the Thirty-six-reed, Keyed Amplified Sheng family, including a treble, alto, tenor, and bass sheng.

The wind chest is a removable structure. Yang Daming of the Central Broadcasting National Orchestra successively designed a twenty-five-reed, thirty-two-reed, and thirty-six-reed keyed, square sheng. In 1970, Mr. Wang Huizhong of the Central Chinese Orchestra designed and manufactured the thirty-six-reed, keyed amplified square sheng, based on Yang's thirty-two-reed keyed sheng, which overcame the disadvantages of the traditional sheng's unbalanced volume between higher and lower pitches, and strengthened its ability to play complex chords and polyphonic music. The instrument structure has the bass pitches in the middle and the treble on both sides, with half steps in succession, from the middle to both sides alternately.

Such a sheng can play chromatic or atonal repertoire, whereas its ability to play traditional harmonies fluently is gone. Wang worked with Sun Rugui, now retired from Beijing National Instruments, and Zhao Bochun, a retired worker, to make a removable wind chest for these square sheng. They cut the wind chest's upper surface off the body and reinstalled it in back with four buckles, making it much easier to disassemble, repair, and clean the sheng. Later, Zheng Dehui, of the Hong Kong Chinese Orchestra, and Zhao Hongliang, of Hebei Hongliang National Instrument Factory, worked on this model to improve the production process and reduce air leakage.

Thirty-six-Reed, Oval-Shaped Sheng with Folded Pipes | Jiang Langchan (蒋朗蟾) and Jiang Wujian (蒋无间)

In 1980, Jiang Langchan and Jiang Wujian completed the development of the Thirty-six-Reed, Oval-Shaped Sheng with Folded Pipes, based on their design for the Twenty-seven-Pipe, Oval-Shaped Sheng. The blow pipe is shorter, and the pitches are arranged in three rows, like the pitch arrangement of a square sheng. Thirteen of the pipes are folded pipes, creating low pitches with the long air column. The folded pipes are shaped like an upside-down U, with one side shorter than the other. A shorter pipe, creating pitches an octave higher, is placed under the shorter side of the U. Folded pipes are controlled with added keys, and short pipes are controlled by finger holes in the traditional manner. The position of the key is placed close enough to a hole on each pipe that the player can easily control two pitches with a single finger. The remaining twenty-three pipes are arranged in the traditional manner, controlled by fingering holes. The range is G–G3, with complete half steps except for F3. Although modulation is easy on this sheng, its heavier weight remains an issue, thus the model is not widely used by players.

Thirty-four-Reed Sheng | *Liu Rongguang (刘荣光)*

In 1980, Liu Rongguang, of the Chongqing Acrobatic Art Troupe, developed the Thirty-four-Reed Sheng, based on the design of the traditional seventeen-reed sheng, and using the same shape as the traditional round sheng. In terms of the pitch arrangement, the square sheng pipes are divided into two parts, front and rear. The front half forms a semicircle of fourteen pipes. Another five pipes are placed within the semicircle, close to the outer pipes. The rear part forms another semicircle, with thirteen pipes and two extra pipes inside the semicircle, which are controlled by added keys placed between pipes. Both finger holes and keys are used for performance. A performer can play pitches by pressing the keys either with fingers or palms. This sheng has a short blow pipe and is supported on the leg with a metal frame when playing. Amplifier tubes are installed on the pipes between F2 and E3. The whole range is G–E3, nearly three octaves with complete half steps. Since playing this model is not as easy as the thirty-six-reed square sheng, it is not very popular.

Twenty-four-Pipe, Thirty-Reed Double-tone Sheng | *Dong Hongsheng (董洪笙) and Sun Rugui (孙汝桂)*

In 1980, basing his design on Zhang Zhiliang's Twenty-four-Reed Sheng, Dong Hongsheng modified the first, ninth, ten, eleven, twelfth, and nineteenth pipes of the outer ring to "dual-tone pipes," thus increasing the pitches available, and creating the Twenty-four-Pipe, Thirty-Reed Double-tone Sheng. This model has a round wind chest. The double-tone pipes were designed by Sun Rugui. A sheng pipe is vertically divided into two halves with a partition. A reed is inserted into each half. The outer reed is controlled by a finger hole, the inner reed controlled by an added key. This sheng in D has a range of A–G3, nearly three octaves. The range from E to D3 has complete half steps, making modulation convenient. The traditional Chinese repertoire requires no such wide range, but the pipes' diameter is limited, making its volume relatively weak. However, the fingering on Dong Hongsheng's design more complicated than the thirty-six-reed square sheng, so again, the model is not popular among players.

Sz-1 Forty-nine-Reed Sheng | Yu Xuehong (于学洪) and Wang Junhua (王俊华)

In 1982, Yu Xuehong and Wang Junhua, of the Tianjin Quyi Troupe, together developed the Sz-1 Forty-nine-Reed Sheng, adopting the concept of folded pipes to reduce the weight of the sheng and the complexity for fingers to reach keys. Pipe holes and added keys on the same pipe combination are placed close to one anothers so one finger can control two tones at once. The bass pitches are controlled by keys installed around the wind chest, and the treble pitches are controlled by finger holes in the traditional manner. Amplifying tubes are installed on every pipe and the keys are made of lightweight materials to reduce the instrument's weight. There is a built-in push-pull frame under the wind chest to adjust the height of the instrument in performance. The sound range covers A–A3, up to four octaves with complete half steps, making modulation convenient. However, due to its large size, and newer sheng repertoire not requiring such a wide range, this sheng model is not widely used.

Twenty-nine-Reed Round Sheng with Removable Blow Pipe | *Zhao Xidao (赵喜道)*

In 1982, Zhao Xidao, of the Art School of Gansu Province, designed the Twenty-nine-Reed Round Sheng with a removable blow pipe and wind chest. This model was based on Mu Shanping's (牟善平) twenty-six-reed sheng. The prototype was made by Sun Jiadong (孙家东), from Beijing, and Hu Zhiguo (胡治国), from Suzhou. Zhao adopted the concept of "parallel-connected pipes." One long and one short closed-end pipes are attached together. The longer pipe is considered the main pipe, the shorter considered the support pipe. The main pipe is closed in the middle, leaving the lower part with a shorter air column to execute the higher pitches. A hole is drilled in the upper part of the pipe to help build a long air column by connecting the main and support pipes to execute lower pitches. A button is added to seal the pitch window, drilled in the upper part of the main pipe and controlled by a levered key. This is a transformation of the idea of folded pipes.

The range is G–F#3, with complete half steps between F#1 and D3. This sheng can play in nine keys and retains the shape, pitch arrangement, and skills of the traditional Seventeen-reed Round sheng. In addition, the designers adopted an S-shaped, long blow pipe and a removable wind chest, making maintenance easier. However, due to its incomplete modulation and poorer performance than the thirty-six-reed square sheng, this sheng is not widely used.

Removable Wind Chest Part | *Wang Huizhong (王慧中), Sun Rugui (孙汝桂), Zhao Bochun (赵伯纯)*

In 1983, Wang Huizhong, Sun Rugui, and Zhao Bochun together developed the removable wind chest for the thirty-six-reed sheng, thus improving the cleaning and drying of reeds. They separated the upper surface of the sheng's wind chest from its main body. A rubber ring was placed in between, and buckles installed to lock the parts together. This also helps reduce the abrasion of pipe's feet during installation and makes for easier cleaning. Nowadays, almost every sheng model has adopted this kind of design.

Thirty-seven-Reed Double Wind Chest Sheng | *Shao Chunliang (邵春良)*

In 1983, Shao Chunliang, from the Sichuan Conservatory of Music, designed the Thirty-seven-Reed Double Wind Chest Sheng. This kind of sheng maintains the general appearance of a traditional sheng, with nearly the same pitch arrangement. The structure basically combines two traditional sheng, one placed in the original position and the other upside down. The upper and lower sheng wind chests are connected by a pipe. All twenty-one pipes are the same length and all pipes are close-ended, except for five pipes producing the lower five pitches. Pipes of the upper sheng are controlled by levered keys, and the pipes of the lower sheng are controlled by traditional finger holes. The keys are placed close to the press holes, making them convenient to operate with the same finger. This sheng has a total of twenty-one pipes and thirty-seven reeds. The range is G–G3, up to three octaves, with complete half steps. However, because the head is heavy and inconvenient to operate, this model also remains unpopular.

Three-Leaf-Shaped Sheng | *Zhou Bingyu (周炳玉)*

In 1986, Zhou Bingyu developed a sheng shaped in the pattern of three leaves with a novel structure based on the traditional round and square sheng. This model is composed of three groups of six pipes, in a small square sheng. The pattern of three leaves features the blow pipe set between the first two. Finger holes are distributed on the outside of the leaves. The player holds the two front leaves with the palms of both hands and uses the thumb and index finger to control the sounds emanating from the first two leaves. The middle and ring fingers are used to control the sound emanating from the third leaf. In total, this model has eighteen pipes with eighteen reeds. The range is F#–B1, with incomplete half steps, thus this model can only modulate in five keys, C, F, G, D, and B♭. An F is missing in the key of B♭. Due to its incomplete half steps, inconvenient mechanisms for modulation, and narrow range, this model is not widely used.

Thirty-Reed Square Sheng | *Huang Linqian (黄林潜)*

In 1988, Huang Linqian developed the Thirty-Reed Square Sheng, based on the Henan square sheng. Three rows—front, middle, and rear—have eight tubes each, of which six pipes are folded pipes. A sheng tube is divided into two halves vertically, and each half is installed with a reed at the lower end like other normal sheng pipes. The outer pipes are controlled by finger holes in the pipe; inner pipes are controlled by added lever keys. The model uses a total of twenty-four pipes and thirty reeds, each pipe equipped with an amplifier

tube. Huang's sheng is also equipped with an electric, auto-controlled and heated wind chest

to prevent condensation on reeds when warm breath hits cold copper reeds during

performance, an S-shaped blow pipe, and an adjustable frame, which helps to sit the sheng

firmly on the thigh. The range is A–A3, up to three octaves, with complete half steps between

F#1 and D3. Because its overall performance is not as good as the Thirty-six-Reed Square

sheng, this model is not widely favored.

Thirty-seven-Reed Keyed Sheng | *Fan Yuanzhu (范元祝)*

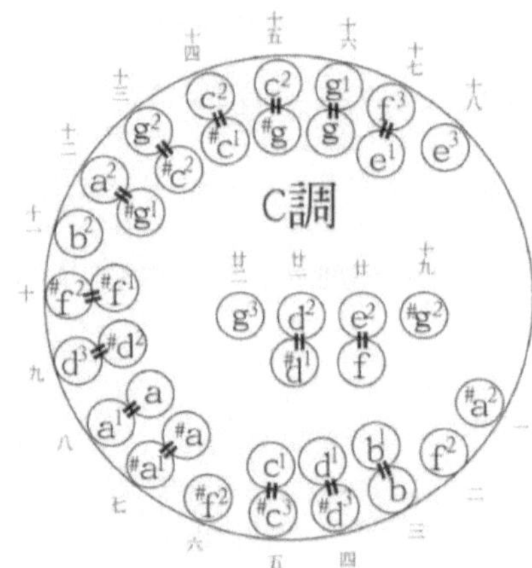

Figure 23. Thirty-seven-Reed Keyed Sheng.[63]

In 1988, Fan Yuanzhu of the Guizhou Institute for Nationalities developed the

Thirty-seven-Reed Keyed Sheng. Adopting the round sheng's pitch-arrangement, the pitches

---

[63] Dongsheng Liu, *Zhongguo Yueqi Tujian* [Chinese Musical Instruments Illustration] (Qingdao: Shandong Education Press, 1992), 156.

of this sheng are arranged in a "G" shape. Eighteen pipes are on the outside, with four pipes inside, thus twenty-two pipes in total. Except for the first, second, sixth, eleventh, eighteenth, nineteenth, and twenty-first pipes, the rest of the Sheng pipes are in the form of "compound pipes." The outside pipes are controlled by finger holes, and inside pipes are controlled by added levered keys. The range is A–A3, up to three octaves, with complete half steps, and modulation is easy. The pipes playing pitches higher than A1 are installed with an amplifier tube. Due to its heavy weight, this model is not conducive to hold and play, so has few users.

Thirty-eight-Reed Sheng and Forty-two-Reed Sheng | Zheng Dehui (郑德慧)

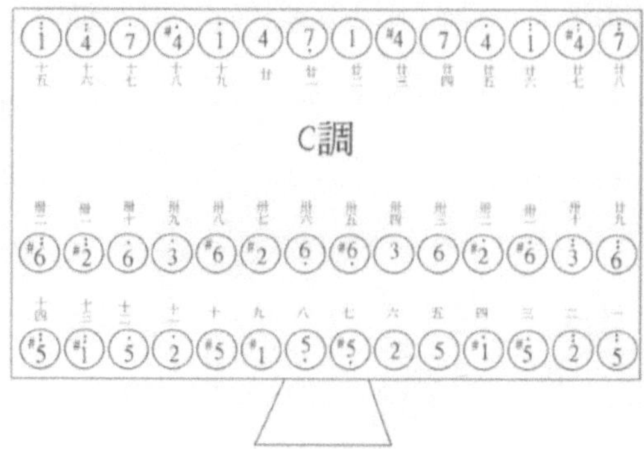

Figure 24. Forty-two-Reed Keyed Sheng.[64]

In 2000, Zheng Dehui, of the Hong Kong Chinese Orchestra, designed the Forty-two-Reed Keyed Sheng, based on Wang Huizhong's Thirty-six-Reed Keyed sheng. Its prototype was made by Zhao Hongliang (赵宏亮) from the Hebei province. The Forty-two-

---

[64] Dongsheng Liu, *Zhongguo Yueqi Tujian* [Chinese Musical Instruments Illustration] (Qingdao: Shandong Education Press, 1992), 156.

Reed sheng added three pipes on both the left and right sides of the thirty-eight-reed sheng, expanding the range to a perfect fourth higher than the highest pitch of the thirty-eight-reed sheng. The new range is G–C4, covering three and a half octaves, with complete half steps. Moreover, Zheng Dehui replaced the nickel-plated copper amplifier tube with wolfram-plated tubes. There are two types of forty-two-reed sheng: Type 1 retains the fingering and sitting performance style of the original thirty-six-reed sheng and Type 2 completely changes the pitch arrangement, and it can be played in both standing and sitting positions.

In the last few decades, musicians and sheng-makers have devoted immense efforts to the development of the instrument, creating numerous types of new sheng. The examination above of these new sheng demonstrates how reform ideas followed six directions summarized in Chapter 2. Some of these sheng have already disappeared, some remain in use today, and some have become the basis of later sheng models.

CHAPTER IV

GLOBALIZATION AND NEW ROLES FOR TRADITIONAL AND MODERN SHENG

PERFORMANCE

Social and Musical Roles of the Traditional Sheng Before 1976

Playing Court Music

In tracing the origin of the piano and its European environment historically, Max Weber concluded that the piano was considered "a significant piece of middle-class furniture" among Nordic peoples.[65] If a musical instrument is closely related to the social identity, or class, of a group, the social class of the group using this instrument can be viewed through the musical instrument. Like the piano's role in Europe, and many other musical instruments, the sheng has been associated with and enhanced certain social identities since its invention.

The Confucian scholars categorize people in ancient China into five occupations—the ruling class (emperors, kings, and feudal lords), the *shi* (gentry scholars), *nong* (peasant farmers), *gong* (artisans and craftsmen), and *shang* (merchants and traders). The ruling class and gentry scholars are considered as upper classes, and the latter three classes are considered as ordinary people. In the history of ancient Chinese musical instruments, according to the classification of Chinese musical instruments, the "金石 jin shi," or instruments like bells and chimes, served the emperors, and the "竽瑟 yu se," or instruments like the sheng, yu, and zither, served the feudal lords. Such instruments are culturally and directly related to the

---

[65] Max Weber, *The Rational and Social Foundations of Music* (New York: Martino Press, 2010), 174.

ruling class. The "琴簫", or *qin xiao*, or *guqin* and *xiao*, served the Chinese literati, and the "瓴缶" or "*ling fou*," i.e., percussion music played on a ceramic or bronzeware jar, served the ordinary people. Initially, the sheng was devised to meet the preferences of the upper class and was played in the court. The ancient Chinese book, "Han Feizi: Nei Chu Shuo," observes an interesting allusion, that might be loosely translated as "pretend to work hard in a team" to everyone in the future.[66]

During the Warring States Period (475–221BCE), Tian Pijiang, the last King of the State of Qi, was fond of listening to yu (large sheng) ensembles. He often requested that 300 yu players form a large ensemble. The king treated his musicians well, and the story goes that a non-musician, Mr. Nanguo, heard of this and managed to become a member of the ensemble. Whenever the band played for the king, Nanguo stood in line and pretended to play. No one listening to the ensemble realized Nangao was making no sounds at all. As a result, he enjoyed the king's favor just as the other musicians. When the king died, his son Tian Sui became the new ruler, who also liked music for the yu. However, he preferred solos, and ordered the yu musicians to play one by one, so Nanguo was run out of the palace. This story not only teaches some timeless truths, but also describes the sheng's use in the court, either as a solo instrument or as part of an ensemble. The sheng ensemble played either as featured ensemble or serving as accompaniment.

---

[66] Fei Han, *Hanfei Zi*, ed. Juanxia Ren, (Beijing: China Textile Press, 2017), 167.

Playing Sacred Occasions

As mentioned in Chapter 2, the sheng spread from royal courts to other levels of society during the unrest, or Warring States period. By the end of the Tang Dynasty (618 CE-907 CE), the sheng was already quite common and popular among lower class people. Because of its popularity, the sheng began to be used for sacred performances, including in religious ensembles for Buddhist, Taoist, or other religious rituals.

In both court music and for religious occasions, sheng players performed within ensembles of other instruments. Because the sheng's role often was collaborator with percussion, such ensembles collectively were referred to as *guchui*, or "wind and percussion" ensembles. According to the leading instrument and styles of repertoire played, the ensembles could be roughly classified into three categories of ensemble leadership: 1) *shengguan*, led by the sheng-player and guanzi, 2) *chuida*, led by the suona, and 3) *daji*, led by percussion instruments. These ensembles overlapped in terms of instrumental make-up and repertoire, but they had individual systems and classification methods for deciding what instrument was featured onstage. In activities related to religious or sacred content, the shengguan ensemble was most important, and sometimes the chuida ensemble also played this music.

For example, when people held sacrifices to ancestors, or festivals, funerals, temple fairs, and other such activities, shengguan ensembles would often appear; a chuida ensemble, led by the suona, rarely appeared on such occasions. Even when the chuida ensemble did appear, the shengguan ensemble occupied the core position, and the chuida group played outside the temple gate. This could be due to the subtle and elegant style of the shengguan, the more complicated technique, and a taste of Zen in the style of music.

The style of the chuida ensemble is more lively and cheerful. Moreover, some sheng music is inherited from court music, so feels more solemn or calm. These three kinds of ensembles sometimes appeared on the same stage in actual performances, but they always performed their own duties in terms of the music. Figure 25 shows the "Illusion of Maitreya-sūtra" mural, on the south wall of Cave No. 25, of Yulin Caves in Guazhou County, illustrating how the band was organized and played in Buddhist ritual occasions in the mid-Tang Dynasty.

Figure 25. Mural on the south wall of Cave No. 25 built in the mid-Tang Dynasty (781-848 CE), Yulin Caves, Guazhou County, Gansu Province.[67]

The "Jingyinhui" of Beijing Zhihua Temple specifies the organization of a shengguan ensemble. The musical instruments on display include the sheng, guanzi, flute, yun gongs, drum, clappers, and cymbals. Moreover, full-time musicians were trained in the

---

[67] "Yulin Cave No. 25." Digital image. Digital Dunhuang. Accessed April 15, 2021. https://www.e-dunhuang.com/cave/10.0001/0001.0002.0025.

temple to play these instruments.[68]

Figure 26. Guchui ensemble from Shangdong Province.[69]

Playing for Folk Activities

Although the shengguan ensembles had certain advantages in sacred activities in

the early days of the Tang Dynasty, as time went by, the lively sound of chuida ensembles,

especially the sound of the suona was preferred by the folk-people and the sheng gradually

lost its lead position in ensembles. An increasing number of suona-led ensembles also took

over music for folk activities. The sheng played more of an accompaniment role because of

its unique ability to play harmonies.

Playing Traditional Opera

In the Liuzi Opera, one of four ancient types of Chinese opera, the sheng plays as

part of the accompaniment ensemble, which is in two parts: *wenchang* and *wuchang*. The

---

[68] Library of Chinese National Academy of Arts, *Zhihuasi Jing Yinyue* [Jing Music of Zhihuasi Temple] (Beijing: Culture and Art Publishing House, 2020), 26.
[69] Ibid. 26.

wenchang part comprises silk and bamboo instruments; the wuchang comprises wind and percussion instruments. The three major instruments of wenchang are dizi, sheng, and *sanxian*, a three-stringed fretless plucked lute. The sheng plays harmonies for the band, the bamboo flute can only play melodies, and the sanxian focuses on improvisations; together they make sufficient textures to offset the voices.

For nearly 3000 years, regardless of the occasion in which it was played, the sheng's role differed in different kinds of ensembles. Either leading or following, the instrument was used in collaboration with other instruments, serving different roles for different occasions. This gave the sheng both advantages and disadvantages. On the positive side the sheng could meet more than the average demands for performance, which also motivated musicians and makers to develop modifications for the instrument. On the negative side, the possibilities for the sheng as a solo instrument were almost completely overlooked. No solo pieces are known for sheng from ancient China. A sheng player could either play tunes adapted from formal court music or play harmonies accompanying a lead instrument's melodies.

From Traditional Ensembles to Conservatory Style

Not until the beginning of the twentieth century did China begin to adopt the western education system, establishing new music institutions, including music-related degrees in universities and specialized music conservatories. At this point, traditional Chinese music education, no longer an oral tradition where masters led apprentices, became an area of academic study. Students with interests in music, who hoped to study further in the field, could thus pursue degrees in performance, composition, theory, and other disciplines. Sheng performance was among these options.

Not surprisingly, the academic education differed from the oral tradition of teaching. Students entering music conservatories, in addition to learning traditional repertoires that had been handed down (and are to this day), they also learned newer pieces composed for sheng. Since then, the role of the sheng has expanded as an academic focus in both performance and historical aspects.

As the performance practice, use, and study of sheng changed, the music written for this instrument changed in the following ways:

1) The music style played by sheng expanded from traditional Chinese styles to fusions of styles from Chinese and western musics. Later, sheng pieces were composed entirely in the western twelve-tone equal temperament tuning. The changes in the style of sheng music unfolded along with changes in the morphology of the instrument, as discussed in Chapter 3. Although the physical shape and construction of the sheng changed between the 1950s to the 1980s, this period comprises only its infancy. During this period, modifications for designs of this instrument still mainly copied the traditional seventeen-pipe round sheng, and it could all be categorized as traditional Chinese sheng. The most representative pieces of this period are Dong Hongde (董洪德) and Hu Tianquan (胡天泉)'s *Fenghuang Zhanchi* [Phoenix Unfolds its Wings] (1956), and *Jindiao* [Tune of Jin] (1960), composed by Yan

Haideng (阎海登). Representative sheng music in this period was composed solely on the basis of traditional tunes.

In the 1980s, the morphology of the sheng developed vigorously, and many ensembles and concertos were composed. At this time, composers modeled their sheng works on western concert works for other instruments, encouraging other composers to explore the sheng's possibilities in a new context. The Chinese polyphonic master, Chen Mingzhi (陈铭志), completed a sheng fugue *Yuan Cao Fu* [Ode to the Prairie] (1988). Zhao Xiaosheng (赵晓声) applied polyphonic techniques and composed *Huan Feng* [Calling for the Phoenix] for the thirty-seven-reed sheng, based on a suona concerto of the same name.

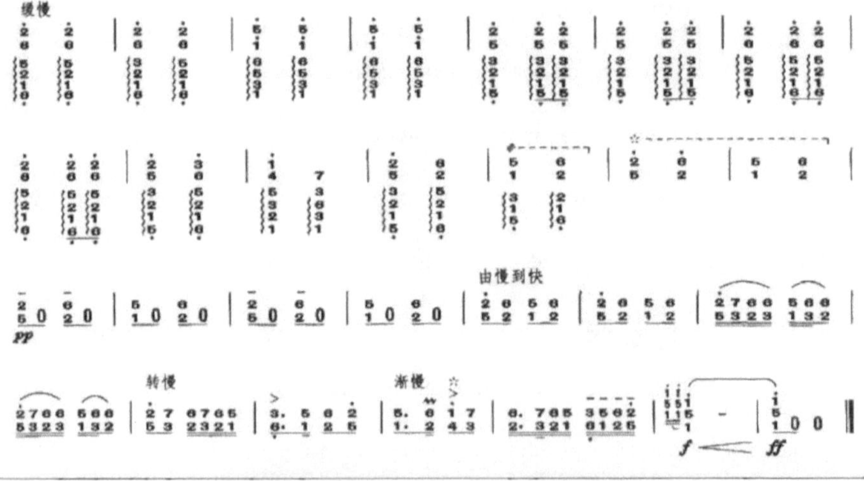

Figure 27. *The Phoenix Unfolds its Wings*, common use of fourths, fifths, octaves, but also thirds.

2) New types of sheng broke with the traditional pentatonic concept of tuning, which treats fourth and fifth chords as core harmonies. New skills were devised for the

tongue and fingers. Nikolai Rimsky-Korsakov's *Flight of the Bumblebee*, transcribed for

sheng, not only demonstrated the progress of the improved sheng, which could now play

semitones, but also showed the performer's breadth and fingering skills on fast performance.

This kind of performance also featured the results of the academic or professional and

systematic training of sheng technique.

Figure 28. Sheng version of Rimsky-Korsakov's *Flight of the Bumblebee*, mm. 1–21.

3) New compositions enhanced the sheng's ability to play harmonies. More

complex harmonies, moving away from the original fourth, fifth, and octave harmonies, such

as thirds, sixths, triads, and sevenths were employed in composition. These colors also

enriched the sound effects and performance techniques of the sheng. Based on modern

harmony and jazz harmony, Jia Guoping (贾国平) composed the trio *Tremors of

Degradation* for sheng. While not easy to play on a traditional sheng, this piece can be

performed on the thirty-six-reed sheng.

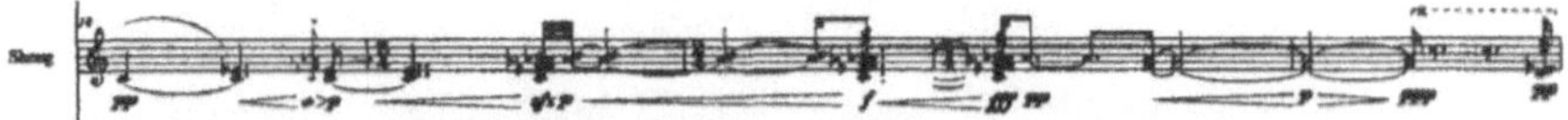

Figure 29. Jazz harmonies used in *Tremors of Degradation*, mm. 10-14.

4) With the advent of a variety of sheng with complete semitones available, the view changed of the sheng as difficult for performing modulations. Musicians began to explore the instrument's ability to play polytonic and atonal works, with complex harmonies outside the general repertoire for sheng, and exclusively works for certain fingerings of a specific kind of sheng. If a player had tried to play such a work with a sheng with little ability to modulate, he could execute the harmonies only with extremely challenging fingering, with some passages being completely unplayable. Representative works of polytonic and atonal sheng music include *Huanfeng*, mentioned above, and *Chuntian De Ge* [Song of the Spring], by Wang Jue (王觉) and Mou Nan (牟楠).

Figure 30. Jue and Nan, polytonic and atonal passage for sheng, *Song of the Spring*, mm. 1–4.

Figure 31. Jue and Nan, polytonic and atonal passage for sheng, *Song of the Spring*, mm. 5–7.

5) The sheng's role was no longer limited to solo or accompaniment in certain types of performances. Audiences began to appreciate its deeper nature and unique sonorities. Not only was it featured in traditional music, as both solo instrument or in ensembles, concertos, and orchestras, but it also became integrated in genres such as electronic music, live stage productions, dance, opera, and theater performances. In the chamber music piece, *Ji V* [Silent V], composed by Qu Xiaosong (瞿小松), the sheng forms a part in a performance art piece. The sheng player walks on stage, as a messenger, slowly moving from one side to the other, with a pondering step. In the neo-visual electronic concert, *Gan. Dong* [Feel. Move], the sheng performers also dance through the entire concert hall in accordance with the melody and rhythm. In Zhong Zhiyue's (钟之岳) multimedia stage work, *Shen. Ren. Chang* [The God. Human. Smooth], in addition to playing the melody, sheng players also use the instrument as a prop, a partner to dance with. Wu Wei, a sheng performer in Germany, plays with a drum set to perform *The Blue Desert*. This kind of collaboration breaks boundaries between nationalities and music genres, and it has met with great success since 2000.

Further Performance Roles for Sheng in the Globalized World

Since the sheng has been taught in conservatories and sheng-makers have designed

different types of sheng to fit equal-temperament tuning, the instrument has become suitable

to play in various types of ensembles worldwide, reaching stages everywhere. In the twenty-

first century, after two decades of preparation after the Chinese Reform and Opening in the

1980s, China entered economic and cultural globalization, integrating in every aspect. Music,

like other cultural products, became a commodity, and sheng-makers and performers sought

to promote this instrument more by improving its quality and adding repertoire to benefit

economically.

As a result, modern sheng players are staking their claims in terms of several new

musical aspects:

1) Expanding collaborations that incorporate the sheng in other forms of music, such

as electronic music, dance, theater, and jazz.

2) Performing musical works created with emerging compositional concepts and

techniques.

3) Joining international music groups and collaborating with traditional musical

instruments from other countries and cultures.

The Germany-based sheng virtuoso, Wu Wei (吴巍), has brought the instrument

onto new stages in the twenty-first century. Born in 1970, in Gaoyou, Jiangsu, Wei pursued

his degree in sheng performance at the Shanghai Conservatory of Music. In 1993, after

graduation, he started a professional career as a sheng soloist, performing with the Chinese

Music Orchestra of Shanghai. In 1995, he won a scholarship from the DAAD (German

Academic Exchange Service) and FNS (Friedrich Naumann Foundation), which brought him

to Berlin, where he is currently located. However, Wu Wei still spends time teaching as a

sheng professor at the Shanghai Conservatory of Music.

Figure 32. Wu Wei (吴巍), sheng virtuoso based in Germany.[70]

Benefiting from the learning experience in both Chinese and western cultures, Wu

Wei became a musician who integrated the ancient Chinese instrument performance tradition

with various types of other musical performances. The infinite possibilities offered by his

sheng performance, in terms of melody, harmony, rhythm, and polyphony, have thus led him

to collaborations with musicians and ensembles in both traditional and orchestral settings, as

well as improvising in solo concerts or with jazz bands. He is well known for playing in a

wide range of music styles, from Baroque to electronic music.

Wu Wei has collaborated with many composers, including: Huang Ruo (黄若)

(*Xuan Huang* [The Color of Yellow], 2007); Guus Janssen (*Four Songs*, 2008); Unsuk Chin (*Su*, 2009), Jukka Tiensuu (*Teoton*, 2015); Bernd Richard Deutsch (*Phaenomena*, 2019); Ondrej Adamek (*Lost Prayer Book*, 2019); Donghong Shin (*Anecdote*, 2019); and Enjott Schneider (*Change*, 2003, and several other concertos). He has performed with many great orchestras and ensembles of our time.[71]

Wu Tong, another significant sheng virtuoso, has brought the sheng to the world-music stage. Born in 1971 in Beijing, Wu Tong began studying sheng and suona with his father at age five. In the late 1970s, he heard some American country music on cassette tape, an experience which would later change his career trajectory. In addition to receiving training on traditional Chinese wind instruments, he also studied different kinds of western music, and became a front man for the groundbreaking rock band, Lunhui [Again], in 1992. In 1994, Wu Tong graduated from the Central Conservatory of Music.

---

[71] Wu Wei has performed with the Berlin Philharmonic, under Kent Nagano, the Seoul Philharmonic, under Myung Whun Chung, the Los Angeles Philharmonic, under Gustavo Dudamel, the BBC Symphony, under Ilan Volkov, the Cabrillo Festival and Sao Paulo Symphony, under Marin Alsop, the Royal Stockholm Philharmonic, the New York Philharmonic, under Susanna Mälkki, the Netherlands Radio Philharmonic, under Jaap van Zweden and Edo de Waart, the Helsinki Philharmonic, under Matthias Pintscher, Holland Baroque, the Ensemble InterContemporain, the Atlas Ensemble, and the NDR Big Band, and with soloists such as Guus Jansen (organ), Wang Li (王利; or jaw harp), and Pascal Contet (accordion).

Figure 33. Wu Tong (吴彤), sheng virtuoso.

Figure 34 Silk Road Ensemble Members

As a musician, Wu Tong has performed as a soloist in symphony orchestras.[72]

After meeting Yo-Yo Ma (马友友) in 2000 at Tanglewood, Wu Tong became a founding

member of the renowned, Grammy-award-winning Silk Road Ensemble and has since

---

[72] Wu Tong has collaborated with New York Philharmonic, the London Sinfonietta, the Chicago Symphony Orchestra, the Hong Kong Philharmonic Orchestra, and the Singapore Symphony Orchestra.

performed sheng on all Silk Road Ensemble recordings, thus playing with tabla, Galician gaita, cajón, and other instruments of the non-European music world. He collaborated with Yo-Yo Ma on *Kuai Le* [Happiness], and was featured on the CD, *Yo-Yo Ma and Friends: Songs of Joy and Peace*, which won the Grammy Award for Best Classical Crossover Album in 2010. In November 2013, Wu Tong joined Yo-Yo Ma in the world premiere of *Duo*, a double concerto written for them by the Chinese composer Zhao Lin (赵琳). Wu Tong's efforts have further expanded the stage for this ancient instrument and enhanced its reputation and instrumental capacities with unprecedented possibilities in the context of globalization.

CHAPTER V

CONCLUSION

In beginning to sort out the history of the sheng, this thesis provided an overview of the aerophone's various ways of use over 3000 years, since its first appearance as the leading instrument in court music, and examined in particular its evolution since 1949, playing either as solo or accompaniment instrument. The sheng has been legendary in that, although its use has fallen in and out of favor according to social and political trends of Chinese music history, its players and makers have always adapted to the times, seizing on new developmental opportunities, and turning the loss of traditions, the political oppression, the pressure from western peers, and other dangers into opportunities for continued survival or growth. After tracing in detail the important morphological developments of the sheng during various periods of Chinese history, including changes in the use and social or political value of musical instruments, and changes in ways of accompaniment styles, its adaptability can be concluded to be the genius of the sheng. In the latest developments of its history there are no fewer than six ways the instrument has been adapted, summarized as follows:

1) The change of the materials for making sheng parts, including the wind chest, pipes, reeds and other parts from gourd, wood, bamboo, and lacquerware to metal and plastic.

2) The introduction of new components, including the Boehm system, keyboards, pedals, and resonating pipes.

3) The change of the instrument's tuning from pentatonic to the twelve-tone equal

temperament system.

4) The introduction of the cooperation of various social forces, including sheng teachers, students, and players with first-hand feedback, artisans with production skills, scholars with knowledge of the development history, scientists with knowledge of physics, acoustics, materials, etc., engineers and manufacturers who can achieve production, and government officials that can help mobilize and integrate resources, etc.

5) The introduction of the concepts of standardization and mass production.

6) The change of the definition of sheng from a single-purpose accompaniment instrument in traditional ensembles to a representative of Chinese musical instruments with multiple roles in different stages, for example, a wind instrument in Chinese national orchestras, a solo instrument in recitals, and a representative of traditional Chinese instruments that can cooperate with instruments from other cultures.

These changes also provided ideas for improvement in over a dozen of new sheng designs after the founding of the People's Republic of China. From the fall of the Qing Dynasty, China's last empire, in 1912, to the establishment of the People's Republic of China in 1949, China experienced an almost uninterrupted 40-year war period. On the one hand, the long-term and high-intensity war severely destroyed the social environment on which traditional Chinese culture depended. When peace finally came, traditional culture had almost disappeared, and people were forced to start from scratch. But during these 40 years (or nearly 100 years if extended to the Opium wars), the ideology of the entire Chinese society had undergone upheavals. Historians have shown that before the First Opium War in 1840, many Chinese people still maintained a mentality of ignorance and arrogance, thinking that their country was a paradise on earth, and the rest of the world was in a state of barbarism and ignorance.

An example was the Emperor Qianlong (1711-1799 CE), who considered the diplomatic relationship between the Qing Dynasty and European countries unnecessary, because these western ambassadors were sent to China "to partake of the benefits of our [China's] civilization."[73] After 1840, China was almost always the loser in foreign wars. In the Second Sino-Japanese War (1937–1945), China suffered its greatest casualties (at least 3,227,926 casualties of the National Revolutionary Army, at least 584,267 casualties of the Eight Route Army, the New Fourth Army, and other regular and militia forces organized by Chinese Communist Party, and 17,000,000-22,000,000 civilian casualties) and property losses (at least 488 billion U.S. dollars at that time) in its history.[74] Certainly, the impact of this blow would be difficult to imagine.

Since the middle of the last century, events and governments in China have caused cultural developments to go toward the other extreme. That is, these have led to sacrifices of its own cultural heritage in order to embrace all that is western. Thus, the Chinese have held

---

[73] Sir. Edmund Backhouse, "Emperor Qianlong's Letter to George III," in *Annals and Memoirs of the Court of Peking*, ed. John Otway Percy Bland, (Boston: Houghton Mifflin, 1914), 322-331.

[74] Micheal Clodfelter, *Warfare and Armed Conflicts: A Statistical Reference to Casualty and Other Figures, 1618-1991*, vol. 2 (Jefferson, North Carolina: McFarland, 1992), 956.

up new standards of quality that prioritized all deemed "scientific" as what China must learn. In philosophy, industry, science, and education to the specifics of these fields, in education the manufacture of musical instruments—their technology and tuning, music pedagogy curriculum, and other details—the Chinese enthusiastically promoted the "westernization" of their culture. Today, whether this adoption of western culture was the best option of the time remains much debated. It has been described as a "life-saving trick," an acceptable option to survive the hard days. In the 40 years between 1949 and 1989, the process of westernizing Chinese music affected many aspects, including the sheng.

Unfortunately, the sheng as the traditional Chinese folk instrument virtually disappeared due to the ravages of years of war, but the changes brought about by the introduction of western music also opened new directions for the sheng. The six developments of the sheng (see chapter 2) reflect the history and differences between western music theories and traditional Chinese music. The materials of the traditional instrument had the problem of being difficult to maintain and carry, so with the development of industrial technology, more types of tubes for resonating tubes, springs for lever keys, and buckles for removable wind chest were devised. These advances not only solved the fragility of the traditional sheng, but also contributed to the richness of its sound.

The continuous changes and innovations in the morphology of the sheng influenced new techniques of composing for the instrument, the enrichment of performance skills, the change of the instrument's social functions, and the change of aesthetic concepts. Thus, the sheng was no longer confined to playing traditional tunes, but played works composed with various techniques, as well as polyphonic and atonal works. The sheng's repertoire now includes not only solo and accompaniment roles for the instrument, but also works for sheng in larger ensembles (e.g., concertos) and chamber ensembles. Sheng players of the traditional Chinese musical instrument now collaborate with instrumentalists from cultures worldwide. These changes in its collaboration and use have propelled its proponents and players toward new opportunities for change. Such developments demonstrate that the changes in sheng music are a response to its social and cultural environment. The changes in conformity within the cultural environment have formed a solid basis for continuing new roles for this versatile instrument.

APPENDIX 1

List of Sheng Artisans

Sun, Binglin 孙炳麟, located in Beijing, he is the father of Sun Rugui. He invented the Turntable Keyed Sheng in 1959.

Sun, Rugui 孙汝桂 (1936—2004), located in Beijing, he was one of the most important sheng artisans in modern China who invented a series of sheng including the twenty-four-reed, keyed amplified sheng, thirty-six-reed, keyed amplified round sheng, the thirty-six-reed sheng family, etc. He was the first artisan to apply resonating pipes to sheng pipes.

Sun, Jiadong 孙家冬 (1962—2020), located in Beijing, he is son of Sun Rugui. He introduced the concept of a removable wind chest in 1980s. He and Zhao Xidao invented the twenty-nine-reed round sheng with removable blow pipe in 1982.

Zhao, Bochun 赵伯纯, located in Zhuozhou, Hebei, Zhao Bochun used to work as a sheng artisan in Beijing National Instrument Factory. Together with his eldest grandson Zhao

Hongliang, Zhao Bochun started his own brand "Zhaojia Sheng" [Sheng of the Zhao Family] after he retired from Beijing in 1980s.

Zhao, Hongliang 赵宏亮 (b.1969—), located in Zhuozhou, Hebei, he is the co-founder and current operator of "Zhaojia Sheng." Cooperating with Wang Huizhong, Zheng Dehui, and many other musicians, Zhao Hongliang created thirty-eight-reed sheng, forty-two-reed sheng, and many other kinds of sheng.

Zhao, Xidao 赵喜道, located in Lanzhou, Gansu, Zhao Xidao focused on creating removable blow pipes and wind chests for different kinds of sheng. He and Sun Jiadong created the twenty-nine-reed round sheng with removable blow pipe in 1982.

Wang, Yuquan 王玉泉 (1902—1965), located in Kaifeng, Henan. He invented the fangsheng [sheng with square-shaped wind chest made] which inspired all later designs of sheng with square-shaped wind chest.

Wang, Fenglin 王凤林, located in Tianjin, he was honored as "The King of Sheng Artisans" due to his skills in making and repairing sheng.

Wang, Yinlan 王印兰, located in Tianjin he is the first son of Wang Fenglin, inherited his father's honor title and worked as a sheng artisan.

Wang, Yinde 王印德, located in Jinan, Shandong. As the second son of Wang Fenglin, he moved to Shandong for new opportunities. He modified the seventeen-reed sheng for Hu Tianquan in 1950s. Hu worked with him and together they created a serious of sheng, including the Bawu sheng.

Wang, Junhua 王俊华, located in Tianjin, he is son of Wang Yinlan. He introduced machines and the idea of mass production into sheng industry.

Wang, Yudong 王育东, located in Tianjin, he is the nephew of Wang Junhua. Wang Yudong is both a professional sheng player and an artisan.

APPENDIX 2

List of Compositions

Daixiang Fengqing 傣乡风情 "Folk Songs of Dai People"

Fenghuang Zhanchi 凤凰展翅 "The Phoenix Unfolds"

Gua Hongdeng 挂红灯 "Hang up the Red Lanterns"

Hong 虹 "Rainbow"

Huanle De Caoyuan 欢乐的草原 "Joy on the Prairie"

Huan Feng 唤凤 "Call for the Phoenix"

Jin Diao 晋调 "Tunes in Jin"

Kongque Kaipin 孔雀开屏 "The Peacock Raises"

Qin Wang Pozhen Yue 秦王破阵乐 "Prince of Qin Breaks up the Enemy's Front"

Shuiku Yinlai Jin Fenghuang 水库引来金凤凰 "Phoenix Comes for the New Dam"

Tian'e 天鹅 "The Swan"

Weishanhu Chuange 微山湖船歌 "Fishing Song in Weishan Lake"

Yangguang Zhaoyao Tashikuergan 阳光照耀塔什库尔干 "A Sunny day in Toxkhürghon"

APPENDIX 3

List of Sheng Composers and Representative Pieces

Cao Jianguo 曹建国 (b.1949—2005) Festival in Tianshan Mountain; The Spring View of a Pasture.

Chin Unsuk 陈银淑 (Korea) (b.1961.7.14—) Su- Concerto for Sheng and Orchestra.

Enjott Schneider (Germany) (b.1950.5.24—) Verändrungen- Concerto for Sheng and Orchestra.

Feng Haiyun 冯海云 (b.1956—) Three Polyphonic Solos for the Thirty-six-reed Sheng.

Gao Pei 高沛 (n.d.) Legend of the "Longing for Husband Cloud"- Concerto for Sheng and Orchestra.

Gao Yang 高扬 (n.d.) Princess Wencheng- Concerto for Sheng and Orchestra, Phoenix Comes for the New Dam.

Guan Naizhong 关迺忠 (b.1939—) Symphonic Picture- The Peacock.

Hu Tianquan 胡天泉 (b.1934—) The Phoenix Unfolds, Happy Songs of the Axi People, Riders on the Pasture.

Liu Wenjin 刘文金 (b.1937.5.5—2013.6.27) Rainbow- Concerto for Alto Sheng Ensemble.

Miao Jing 苗晶 (n.d.) Hang up the Red Lanterns.

Mu Shanping 牟善平 (b.1942—) Fishing song in Weishan Lake, Hang up the Red Lanterns.

Wang Qingchen 王庆琛 (b.1938—) Phoenix Comes for the New Dam.

Wang Huizhong 王慧中 (b.1942—) Folk Songs of Dai People.

Wang Huiyi 王会义 (n.d.) Red Flowers Blossom Everywhere.

Weng Zhenfa 翁镇发 (n.d.) Toast Song, The Song of Spring.

Xiao Jiang 肖江 (b.1942.6.20—) In the Silent Night.

Xu Chaoming 徐超铭 (b.1942.12—) Linka's Night, Celebrate the Harvest.

Yan Haideng 阎海登 (b.1930—2004) Tunes in Jin, Folk Song of Yimeng, The Peacock Raises.

Zhao Xiaosheng 赵晓生 (b.1945—) Call for the Phoenix.

Zhang Zhiliang 张之良 (b.1940.1—) Night in the Village, The Winter Hunt, Tale of the White Snake, Joy on the Prairie, Prince of Qin Breaks up the Enemy's Front.

www.ingramcontent.com/pod-product-compliance
Lightning Source LLC
LaVergne TN
LVHW041714190726
843493LV00007B/2086